Lecture Notes in Economics and Mathematical Systems

Managing Editors: M. Beckmann and H. P. Künzi

Systems Theory

110

Charlotte Striebel

Optimal Control of Discrete Time Stochastic Systems

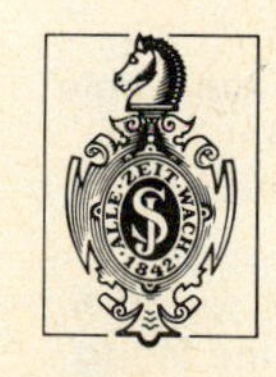

Springer-Verlag
Berlin · Heidelberg · New York 1975

Author

Dr. Charlotte Striebel
University of Minnesota
School of Mathematics
Minneapolis, Minnesota 55455
USA

Library of Congress Cataloging in Publication Data

Striebel, Charlotte, 1929-
Optimal control of discrete time stochastic systems.

(Systems theory) (Lecture notes in economics and mathematical systems ; 110)
Bibliography: p.
Includes index.
1. Control theory. 2. Mathematical optimization. 3. Discrete-time systems. I. Title. II. Series. III. Series: Lecture notes in economics and mathematical systems ; 110.
QA402.3.S84 1975 629.8'312 75-19386

AMS Subject Classifications (1970): 90D15, 93C55, 93E05, 93E99, 94A25

ISBN 3-540-07181-4 Springer-Verlag Berlin · Heidelberg · New York
ISBN 0-387-07181-4 Springer-Verlag New York · Heidelberg · Berlin

Printed in Germany
Offsetdruck: Julius Beltz, Hemsbach/Bergstr.

TABLE OF CONTENTS

Chapter 1 - Introduction and Formulation of the Model

1.1 Introduction and Summary

The theory of discrete time stochastic control systems has received impetus from two directions. First, the area, which is followed here in the matter of terminology and notation, is that of engineering applications; and second is the area of sequential statistical decision theory. Though, in their most general formulations, the models used in the two areas are equivalent, the motivation and emphasis is often quite different.

In the area of engineering applications the discrete time stochastic system is usually looked upon as an approximation to a continuous time stochastic system, the sample spaces of the random variables involved are usually Euclidean spaces, the assumptions of the model are based on physical fact and are usually believed quite literally, and the emphasis is on the mechanization and evaluation of optimal or sub-optimal control laws (policy, plan, strategy). A discussion of the motivation and some elementary examples are given by Aoki [1] , Chapter I .

In statistical decision theory the sample spaces, especially the action (or control) space, tend to be finite or countable; motivation comes primarily from models for gambling and economic phenomena; and continuous time has little interest. The sequential statistical decision problem takes the form of a control problem only when an a priori Bayes distribution is assigned to the parameter space. Thus the control problem is seen as a classical decision theory problem where the particular Bayes distribution is viewed with the usual skepticism. Indeed, often many aspects of the model are arbitrary and not to taken literally. As a result emphasis is on existence and properties (stationary, Markov, etc.) of classes of optimal laws (strategies) rather than on explicit construction.

This monograph is intended as a self-contained, but not a comprehensive treatment of discrete time stochastic control systems. Relationships with other work will be pointed out (primarily in the introductory sections), but all essential results will be proved directly without appeal to the literature of stochastic control theory. This is done partly in a desire to present a unified theory and partly because differences in formulation of models (often inconsequential) makes application of other work awkward, requiring elaborate and irrelevant comparison of models and notation. Choice of the model can only be defended on the aribitrary grounds that, to the author, it seems to be the most "natural" and convenient for application to problems of interest. It must be admitted that much of the motivation comes from a desire to generalize to the continuous time problem. The model used here admits generalization to continuous time far more readily than does the decision theory (dynamic programming) model. Much of the appeal of the conditional loss functional (Definition 4.2.3) over more convential loss criteria lies in the ease with which it generalizes to the continuous time situation (see, Striebel [18]).

The principal aim throughout is to develop workable algorithms for the construction of optimal control laws. Hence the emphasis is on sufficient conditions for optimality. In the interest of completeness non-constructive existence proofs are occasionally given, but they are essentially peripheral to the main development. In line with the emphasis on constructibility only the finite time horizon case is considered. While many of the existence results can be generalized to the infinite time horizon case, except for stationary models they lose their constructibility in any practical sense. The stationary model is not considered here at all.

The model treated here (essentially that of Striebel [17] and Aoki [1] p. 21) differs in several details from the dynamic programming model (Blackwell [2] , Strauch [16] , and Hinderer [8]) and the model for controlled random sequences (Dynkin [7] and Sirjaev [15]) . The Dynkin model for the incomplete information problem is defined by specifying certain stochastic kernels (regular conditional probabilities) which together with a control law determine a distribution on the sample space. In section 2.2 it is shown that the kernels P_t defined from the model formulated in section 1.2 perform this same function. Thus the two models are essentially equivalent. Reduction of the incomplete information model to the complete information model requires the existence of the filtering kernels G_t of Theorems 2.2.1-2.2.3 . This result presents few difficulties for countable spaces and is given by Dynkin [7] and Hinderer [8] in this case. Hinderer [8] , section 7, discusses the use of these kernels in the reduction of the problem to the complete information problem. This reduction, which is discussed in more detail in section 2.1 , is implied whenever dynamic programming results are quoted or compared with results of the present work. Chapter 2 is devoted to the distribution theory required to deal with incompleteness of information.

Another difference from the controlled random sequences and the dynamic programming models is in the matter of the loss function. In these models a reward function is specified, and expected rewards are maximized instead of minimizing expected losses. A simple sign change is all that is required to convert one into the other. All loss and cost functions used here are required to be non-negative and possibly infinite. This corresponds to the essentially negative (EN) case of Hinderer [8] and Strauch [16] .

Another inconsequential difference is in the consideration of randomized control laws. It is extremely difficult to justify the use of randomized control laws in engineering applications, and they will not be considered here. However, this makes for no real differences in the theory since in the EN case randomized procedures are not required (see Hinderer [8] Theorems 15.2 and 15.4). Thus the theory is conceptually (though not practically) simplified in that randomized laws are not introduced only to be later discarded.

Markov laws (plans) considered by Blackwell [2] , Strauch [16] , and Hinderer [8] are not appropriate in the incomplete information case. However, in the presence of a sufficient statistic the question of existence of an optimal Markov law and the existence of an optimal law in the statistic are essentially equivalent. Most of Chapter 5 is devoted to the latter problem. This point is considered more fully in the discussion of sufficient statistics in section 3.1.

One last difference from the models of the literature is in the use of conditional expectations. Here they will always be used in the standarad probabalistic sense (e.g. Loève [12] 24.2) without the assumption of regular conditional distributions. A discussion of this difference and its implications is given in section 4.4.

Many properties of stochastic control systems will be isolated and defined. As a general principle, arbitrary spaces, classes of laws, etc. will be considered at the outset; additional properties will be specified and assumed explicitly as they are required. Thus, the model is formulated for arbitrary sample and control spaces though many of the principal results are proved only for Euclidean spaces. Results will rarely be stated for countable or finite spaces. It is usual in the literature of the subject to consider as admissible all control laws which have measurability properties necessary to determine a probability measure on the sample space. Here a class of admissible laws will be specified as part of the model and properties of this class will be introduced as they are required.

No topological assumptions will be made. This results in an emphasis on ϵ-optimal control laws since no general results on existence of optimal laws are possible. Existence of optimal laws is treated from a constructive point of view in that methods are given which produce optimal laws if such exist and otherwise give ϵ-optimal laws.

The linear Gaussian model, defined in section 1.3, provides applications and motivation for much of the theory of succeeding chapters. For example, the Kalman filter derived in section 2.4 using the theory of section 2.3 is the historical forerunner of the general estimation theory of Chapter 2. The properties of the sufficient statistic $\hat{a}_t$ defined for the linear Gaussian model in section 3.4 supplies the primary motivation for the theory of sufficient statistics in Chapter 3. Chapters 6 and 7 provide examples of the application of the optimal control theory of Chapter 5 to the linear Gaussian model.

Chapters 2 and 3 are devoted to distribution theory which is motivated by, but is theoretically independent of, the control problem. Only the "classical information pattern" of Witsenhausen [20] is considered. The definition of a sufficient statistic given in Chapter 3 is slightly different from that of Hinderer [8] , Dynkin [7] , and Sirjaev [15] (see, discussion in section 3.1). The emphasis in Chapter 3 is on methods for determining sufficient statistics for a particular loss structure. The motivating example is given in section 3.1 , again for the linear Gaussian model. The theory of Chapters 2 and 3 depends heavily on appendix section A.1 which is a collection of standard results on stochastic kernels not conveniently available elsewhere. The only original result of section A.1 is Lemma A.1.9 which is crucial for the general estimation theorem (Theorem 2.2.1).

Perhaps the most innovative results are in Chapter 4 where a new definition for the minimal conditional loss functional (Definition 4.2.3) is proposed. In a sense, this definition interprets the previous control law as a part of the "observations" available at time t . While this notion has been part of the folklore of stochastic control theory for some time, it has not been exploited in a rigorous way. Some of the martingale properties of the conditional loss functional developed in section 4.2 are related to properties of excessive functions obtained by Dubins and Savage [5]. A more detailed discussion of the definition and properties of the conditional loss functional that are basic to the method is given in section 4.1.

Chapter 4 depends on appendix section A.2 for the definition and properties of the P-ess **inf** of a family of non-negative measurable functions. Section A.2 depends in turn on the definition of the infimum of a family of finite measures (Dunford and Schwartz [6] , Ch. III 7.5 and 7.6) . All the results of section A.2 are straightforward except for Theorems A.2.2 and A.2.3 which have a bit more depth and are basic to the sub-martingale property of the conditional loss functional.

In Theorems 4.3.5-4.3.7 , the notions of the P-ess **inf** and the conditional loss functional are used to give a construction algorithm that is similar in spirit to the standard dynamic programming method.

Definitions of conditional optimality are given in section 4.3. These are compared with the definitions of Hinderer [8] , Blackwell [2] , and Strauch [16] in section 4.1. A type of optimality (Definitions 5.5.2 and 5.5.3) made possible by the use of sufficient statistics is discussed in section 5.1.

In Chapter 5 a selection theorem of Brown and Purves [3] (stated here as Lemma 5.4.1) is exploited to obtain constructive existence theorems. While the most widely applicable results on the single strong selection class (Definition 5.5.4) can be obtained more directly as in Aoki [1] and Sirjaev [15] , the more systematic development involving the intermediate ideas of selection classes (Definition 5.2.3) , complete selection classes (Definition 5.4.1) and the single selection class (Definition 5.5.1) provides a greater variety of results.

A few comments about notation are in order. First, t is used as a sub and superscript to indicate discrete time. Thus

$$t = 0,1,2,\ldots$$

violatesthe usual convention that t be used to indicate only continuous time. Since continuous time is never used here, there is no chance of confusion. This frees the usual discrete indices such as $i,j,$ and n for other purposes (components of vectors, etc.). Aoki [1] is followed in the use of superscripts to indicate accumulated data, see (1.2.6). Much of the notation used here represents a reaction to the rather Spartan notation of Blackwell [2] (followed also by Strauch [16] and Hinderer [8]) . Time sub and superscripts and explicit arguments of functions, functionals, and stochastic kernels are used perhaps too liberally. Redundancy of notation is used quite unabashedly when it seems helpful in emphasizing a points. Terms will be underlined at the point in the text at which they are defined.

1.2 General Model and Notation

Three interdependent random time series are required to describe the general discrete parameter control problem. The <u>system process</u> x_t models the phenomena that is of primary concern-- the values which are to be controlled. It will be seen later that the system process may also contain components in which there is no real interest, but which are required so that the system process will satisfy the required Markov-like property (1.2.2). The second process z_t is the <u>observation process</u> and consists of those components which can be observed and hence contain the information which is to be the basis for control. The final process is the <u>control process</u> u_t which includes as components values which can be manipulated-- those values which may be adjusted, possibly subject to some constraints.

Loosely stated, the control problem consists of selecting the control variable based on the available information at each step in such a way as to obtain some desired effect on the system process with minimum cost.

It is assumed that the system process satisfies

$$x_0 = w_0 \tag{1.2.1}$$

$$x_{t+1} = \varphi_t(x_t, u_t, w_{t+1}) \quad t = 0,1,2,\ldots,T-1 \tag{1.2.2}$$

and the observation process satisfies

$$z_t = \psi_t(x_t, u_{t-1}, e_t) \quad t = 1,2,\ldots,T \tag{1.2.3}$$

where $\varphi_t(\cdot,\cdot,\cdot)$ and $\psi_t(\cdot,\cdot,\cdot)$ are measurable functions and

$$\omega = \{w_0, e_1, w_1, e_2, \ldots, w_{T-1}, e_T\} \tag{1.2.4}$$

is a vector of independent random variables. The variables e_t and w_t will be called the <u>observation</u> and <u>mechanization</u> errors. The <u>final time</u> T is assumed to be finite.

The distribution of the errors is determined by a product probability space

(1.2.5) $\quad (\Omega, \mathcal{B}_\Omega, P_\Omega)$

where $\omega \in \Omega$ and ω is given by (1.2.4).

Let z^t indicate the data available at time t,

(1.2.6) $\quad z^t = (z_1, \ldots, z_t)$.

Then the control process satisfies

(1.2.7) $\quad u_t = u_t(z^t) \qquad t = 1,2,\ldots,T-1$

where $u_t(\cdot)$ is a measurable function, called the <u>control</u> <u>function</u> at time t. At time $t = 0$, no observations are available so that u_0 is a deterministic constant. This will occasionally be denoted by $u_0(z^0)$ where z^0 indicates no data. Consistent with this, the Borel σ-field generated by z^0 is the trivial σ-field,

(1.2.8) $\quad \mathcal{B}_Z^0 = \mathcal{B}(z^0) = \{\emptyset, \Omega\}$

The spaces of admissible values of x_t, z_t, and u_t will be indicated by $\mathcal{X}_t$, $t = 0,1,2,\ldots,T$, Z_t, $t = 1,2,\ldots,T$ and $\mathcal{U}_t$, $t = 0,1,2,\ldots,T-1$. It will be assumed that $(\mathcal{X}_0, \mathcal{B}_{\mathcal{X}_0})$, $(\mathcal{X}_t, \mathcal{B}_{\mathcal{X}_t})$, $(Z_t, \mathcal{B}_{Z_t})$, $(\mathcal{U}_{t-1}, \mathcal{B}_{\mathcal{U}_{t-1}})$ $t = 1,2,\ldots,T$ are measurable spaces. In most applications these spaces will be finite dimensional Euclidean space identical in t. However, when such assumptions are required they will be explicitly stated. Following (1.2.6) the space of values of z^t, will be indicated by $Z^t = Z_1 \times Z_2 \times \cdots \times Z_t$ with σ-field $\mathcal{B}_Z^t = \mathcal{B}_{Z_1} \times \mathcal{B}_{Z_2} \times \cdots \times \mathcal{B}_{Z_t}$. The measurable spaces $(\mathcal{X}^t, \mathcal{B}_{\mathcal{X}}^t)$ and $(\mathcal{U}^t, \mathcal{B}_{\mathcal{U}}^t)$ are defined similarly by $\mathcal{X}^t = \mathcal{X}_0 \times \mathcal{X}_1 \times \cdots \times \mathcal{X}_t$ and $\mathcal{U}^t = \mathcal{U}_0 \times \mathcal{U}_1 \times \cdots \times \mathcal{U}_t$. Points in $\mathcal{X}^t$ and $\mathcal{U}^t$ will be denoted by $x^t = (x_0, x_1, \ldots, x_t)$ and $u^t = (u_0, u_1, \ldots, u_t)$. The spaces of the random variables w_t and e_t will be denotes by $\mathcal{W}_t$ and $\mathcal{E}_t$.

The measurable functions φ_t , ψ_t together with the spaces $\mathcal{X}_t$, Z_t , $\mathcal{U}_t$ and the probability space $(\Omega,\mathcal{B}_\Omega,P)$ describe a <u>control</u> <u>model</u> and are considered to be given and known. The control functions $u_t(\cdot)$ are to be determined. A sequence of control functions

(1.2.9) $$\{u\} = \{u_0,u_1(\cdot),\ldots,u_{T-1}(\cdot)\}$$

is called a <u>control law</u>. <u>Truncated control laws</u> will be indicated by

(1.2.10) $$\{u^t\} = \{u_0,u_1(\cdot),\ldots,u_t(\cdot)\} \quad .$$

Thus

(1.2.11) $$\{u^{T-1}\} = \{u\} \quad .$$

<u>Equality of truncated control laws</u>, written

$$\{u^t\} = \{\bar{u}^t\}$$

is defined by

$$u_\tau(z^\tau) = \bar{u}_\tau(z^\tau)$$

for all $z^\tau \in Z^\tau$ and $\tau = 0,1,2,\ldots,t$.

A truncated control law defines a function from Z^t to $\mathcal{U}^t$. This function will be denoted by

(1.2.12) $$u^t(z^t) = (u_0,u_1(z^1),\ldots,u_t(z^t)) \quad .$$

When the notation z^τ is used as in (1.2.12) with several different superscripts τ the intention is that z^τ is a contraction of the same vector z^T . Thus for

$$z^t = (z_1,z_2,\ldots,z_t)$$

given or implied and $\tau < t$

$$z^\tau = (z_1,z_2,\ldots,z_\tau)$$

and

$$z^{\tau+1} = (z^\tau,z_{\tau+1}) = (z_1 z_2,\ldots,z_\tau,z_{\tau+1}) \ .$$

For a given control law $\{u\}$, by successive substitution into (1.2.2), (1.2.3) , and (1.2.7) the processes u_t , x_t and z_t can be written as random variables on the probability space (1.2.5) .

$$
\begin{aligned}
x_0 &= w_0 \\
x_1 &= \varphi_0(w_0,u_0,w_1) = X_1^{\{u^0\}}(w^1) \\
z_1 &= \psi_1(x_1,u_0,e_1) = \psi_1(\varphi_0(w_0,u_0,w_1),u_0,e_1) = Z_1^{\{u^0\}}(e^1,w^1) \\
u_1 &= u_1(z^1) = u_1(z_1) = u_1(\psi_1(\varphi_0(w_0,u_0,w_1),u_0,e_1) = U_1^{\{u^0\}}(e^1,w^1) \\
&\vdots \\
u_{\tau-1} &= u_{\tau-1}(z^{\tau-1}) = u_{\tau-1}(Z_1^{\{u\}}(e^1,w^1),\ldots,Z_{\tau-1}^{\{u\}}(e^{\tau-1},w^{\tau-1})) = \\
&\qquad U_{\tau-1}^{\{u^{\tau-1}\}}(e^{\tau-1},w^{\tau-1}) \\
x_\tau &= \varphi_{\tau-1}(X_{\tau-1}^{\{u\}}(e^{\tau-2},w^{\tau-1}),U_{\tau-1}^{\{u\}}(e^{\tau-1},w^{\tau-1}),w_t) = X_\tau^{\{u^{\tau-1}\}}(e^{\tau-1},w^\tau) \\
z_\tau &= \psi_\tau(X_\tau^{\{u\}}(e^{\tau-1},w^\tau),U_{\tau-1}^{\{u\}}(e^{\tau-1},w^{\tau-1}),e_\tau) = Z_\tau^{\{u^{\tau-1}\}}(e^\tau,w^\tau) \\
&\vdots
\end{aligned}
\tag{1.2.13}
$$

As noted before z_0 is fictitious (no data), and u_0 is deterministic. The functions U_τ , X_τ , Z_τ are obviously measurable. The probability measure induced on the space

$$
(\mathcal{X}^t \times Z^t \times \mathcal{U}^{t-1}, \mathcal{B}_{\mathcal{X}}^t \times \mathcal{B}_Z^t \times \mathcal{B}_{\mathcal{U}}^{t-1}) \tag{1.2.14}
$$

by $x_0 = w_0$ and the functions (1.2.13) for $\tau = 1,2,\ldots,t$ will be indicated by $P_{\{u^{t-1}\}}$. For $t = T$, the time index will be dropped. Thus

$$
P_{\{u\}} = P_{\{u^{T-1}\}} \tag{1.2.15}
$$

is a probability measure on (1.2.14) with $t = T$. Expectations with respect to the measure $P_{\{u^{t-1}\}}$ will be indicated by $E_{\{u^{t-1}\}}$. For example,

$$
E_{\{u\}}[\ell(x_t)] = \int_{\mathcal{X}_t} \ell(x_t) P_{\{u\}}[dx_t] = \int_\Omega \ell(X_t^{\{u\}}(\omega)) P_\Omega(d\omega) \ . \tag{1.2.16}
$$

Integration with respect to the measure $P_{\{u\}}$ is indicated by $P_{\{u\}}[dx_t]$ rather than $dP_{\{u\}}$. This convention is particularly convenient when many variables are involved.

The model is used principally to define the conditional probabilities

$$P_{\{u\}}[x_{t+1} \in A \, , \, z_{t+1} \in B | x_t, z^t] \tag{1.2.17}$$

where $A \in \mathcal{B}_{\mathcal{X}_{t+1}}$ and $B \in \mathcal{B}_{Z_{t+1}}$. This is a conditional probability in the usual sense (see, e.g. Loève [12] , p. 341) on the basic probability space

$$(\mathcal{X}^T \times Z^T \times \mathcal{U}^{T-1} \, , \, \mathcal{B}_{\mathcal{X}}^T \times \mathcal{B}_Z^T \times \mathcal{B}_{\mathcal{U}}^{T-1}, P_{\{u\}}) \tag{1.2.18}$$

with respect to the σ-field generated by the random variables (x_t, z^t) (that is, cylinder sets with bases in $\mathcal{B}_{\mathcal{X}_t} \times \mathcal{B}_Z^t$) . However, conditional expectations of this type will always be taken to be functions on the range space of the conditioning random variables rather than on the underlying probability space (see e.g. Lehmann [11], p.39). This convention will be violated only in the case of the fictitious variable z^0 , where by convention consistent with (1.2.8)

$$E_{\{u\}}[L | z^0] = E_{\{u\}}[L] \; . \tag{1.2.19}$$

For the control problem the distributions of z^t are of particular interest. The notation $P_{\{u\}}^t$ will be used to indicate the marginal distribution of $P_{\{u^{t-1}\}}$ on Z^t . Thus for $B \in \mathcal{B}_Z^t$

$$P_{\{u\}}^t(B) = P_{\{u^{t-1}\}}(\mathcal{X}^t \times B \times \mathcal{U}^{t-1}) \tag{1.2.20}$$

defines a probability space

$$(Z^t, \mathcal{B}_Z^t, P_{\{u\}}^t) \qquad t = 1,2,\ldots,T \quad . \tag{1.2.21}$$

It should be noted that $P_{\{u\}}^t$ depends only on the truncated control law $\{u^{t-1}\}$. Similarly, it may be noted from (1.2.13) that the conditional probability (1.2.17) can also be computed with respect to $P_{\{u^t\}}$; that is,

$$P_{\{u\}}[x_{t+1} \in A, z_{t+1} \in B | x_t, z^t] = P_{\{u^t\}}[x_{t+1} \in A, z_{t+1} \in B | x_t, z^t] \quad \text{a.s. } P_{\{u^{t-1}\}} \tag{1.2.22}$$

Parentheses will be used for stochastic kernels (Definition A.1.1) and square brackets will be used to indicate conditional probabilities as in (1.2.22).

Success in the estimation problem and the control problem depends in large measure on the ability to give explicit and simple representations of the dependence of distributions $P_{\{u\}}$ on the control law $\{u\}$. Three stochastic kernels defined from the model will be crucial in this connection. They are

(1.2.23) $$P_t(x_t,u_t;A \times B) = P_\Omega[(w_{t+1},e_{t+1}) | \varphi_t(x_t,u_t,w_{t+1}) \in A, \psi_{t+1}(\varphi_t(x_t,u_t,w_{t+1}),u_t,e_{t+1}) \in B] \quad A \in \mathcal{B}_{\mathcal{X}_{t+1}}, B \in \mathcal{B}_{Z_{t+1}}$$

(1.2.24) $$Q_t(x_t,u_t;A) = P_\Omega[\{w_{t+1} | \varphi_t(x_t,u_t,w_{t+1}) \in A\}] \qquad A \in \mathcal{B}_{\mathcal{X}_{t+1}}$$

(1.2.25) $$H_{t+1}(x_{t+1},u_t;B) = P_\Omega[\{e_{t+1} | \psi_{t+1}(x_{t+1},u_{t+1},e_{t+1}) \in B\}] \qquad B \in \mathcal{B}_{Z_{t+1}}$$

It will be shows in Lemmas 2.2.1 and 2.2.2 that P_t, Q_t and H_{t+1} defined by (1.2.23)-(1.2.25) are stochastic kernels and that they satisfy

(1.2.26) $$P_{\{u\}}[x_{t+1} \in A, z_{t+1} \in B | x_t, z^t] = P_t(x_t,u_t(z^t);A \times B) \qquad \text{a.s. } P_{\{u\}}$$

(1.2.27) $$P_{\{u\}}[x_{t+1} \in A | x_t, z^t] = Q_t(x_t,u_t(z^t);A) \qquad \text{a.s. } P_{\{u\}}$$

and

(1.2.28) $$P_{\{u\}}[z_{t+1} \in B | x_{t+1}, x^t, z^t] = H_{t+1}(x_{t+1},u_t(z^t);A) \qquad \text{a.s. } P_{\{u\}}$$

for all control laws $\{u\}$.

The initial distribution of $x_0 = w_0$ does not depend on the control law and will be denoted by

(1.2.29) $$G_0(A) = P_\Omega[x_0 \in A] = P_{\{u\}}[x_0 \in A]$$

for $A \in \mathcal{B}_{\mathcal{X}_t}$ and all $\{u\}$.

An alternative formulation of a stochastic control model is obtained by introducing directly the stochastic kernels Q_t and H_t and the initial distribution G_0 to replace the structure (1.2.1)-(1.2.5) (see Dynkin [7]).

The general model for a stochastic control system is completed by the specification of a <u>class</u> $\mathcal{D}$ <u>of admissible control laws</u> and the definition of a loss function. It will be said that the law $\{u\}$ is <u>admissible</u> provided $\{u\} \in \mathcal{D}$. The general <u>loss function</u>

$$L(x^T, z^T, u^{T-1}) \tag{1.2.30}$$

is a non-negative, measurable, possibly infinite function on the space (1.2.18). The <u>control problem</u> then consists of selecting a control law $\{\hat{u}\} \in \mathcal{D}$ (if possible) that satisfies

$$E_{\{\hat{u}\}}[L] \leq E_{\{u\}}[L] \tag{1.2.31}$$

for all $\{u\} \in \mathcal{D}$. Conditional optimality will also be defined and discussed in later chapters. While some of the general optimality results of Chapter 4 hold for a general control L, constructive results are obtained only for loss functions of the form

$$L = \sum_{t=1}^{T} c_t(x_t, u^{t-1}) . \tag{1.2.32}$$

The <u>cost functions</u> c_t are assumed to be non-negative, possibly infinite and measurable.

The term control model is used to denote the functions φ_t and ψ_t and the probability space (1.2.5) that satisfy (1.2.1)-(1.2.4). A model together with a loss function L and a class of admissible laws $\mathcal{D}$ is called a <u>control system</u>. The final time T is properly a measurability property of the loss function and hence is included in the control system rather than the control model. However, since it may be inconvenient to define the processes x_t and z_t beyond $t = T$, the final time will usually be treated as part of the model.

1.3 Linear Gaussian Model

A special case of the model (1.2.1)-(1.2.5) will be studied in detail. The system and observation processes are assumed to satisfy

(1.3.1) $$x_{t+1} = \Phi_t x_t + \Lambda_t u_t + w_{t+1} \qquad t = 0,1,2,\dots$$

(1.3.2) $$z_t = \Psi_t x_t + e_t \qquad t = 1,2,\dots$$

(1.3.3) $$x_0 \sim N(\hat{x}_0, \Pi_0)$$

(1.3.4) $$e_t \sim N(0, R_t) \qquad t = 1,2,\dots \quad .$$

(1.3.5) $$w_t \sim N(0, \Gamma_t) \qquad t = 1,2,\dots \quad .$$

It is, of course, assumed that the random variables (1.2.4) are independent. The notation (1.3.3) indicates that x_0 is a Gaussian random vector with mean value vector $\hat{x}_0$ and covariance matrix Π_0 . All vectors are column vectors. The vectors x_t, w_t are assumed to be n-dimensional; z_t, e_t are m-dimensional, u_t is s-dimensional. Thus Φ_t , Π_0 , and Γ_t are $n \times n$ matrices; R_t is $m \times m$; Λ_t is $n \times s$; and Ψ_t is $m \times n$.

No special assumptions are imposed on the control functions (1.2.7). Thus, although we say that (1.3.1)-(1.3.5) defines a <u>linear Gaussian model</u>, the processes x_t and z_t will not in general be normally distributed.

From (1.3.2) and (1.3.1)

(1.3.6) $$z_{t+1} = \Psi_{t+1} \Phi_t x_t + \Psi_{t+1} \Lambda_t u_t + \Psi_{t+1} w_{t+1} + e_{t+1} \quad .$$

Under the assumptions of the linear Gaussian model the stochastic kernels P_t , Q_t , and H_{t+1} defined by (1.2.23)-(1.2.25) can easily be seen to satisfy

(1.3.7) $$P_t(x_t, u_t; x_{t+1} \in A, z_{t+1} \in B) \sim N\left[\begin{pmatrix} \Phi_t x_t + \Lambda_t u_t \\ \Psi_{t+1}\Phi_t x_t + \Psi_{t+1}\Lambda_t u_t \end{pmatrix}, \begin{pmatrix} \Gamma_{t+1} & \Gamma_{t+1}\Psi'_{t+1} \\ \Psi_{t+1}\Gamma_{t+1} & \Psi_{t+1}\Gamma_{t+1}\Psi'_{t+1} + R_{t+1} \end{pmatrix}\right],$$

(1.3.8) $$Q_t(x_t, u_t; x_{t+1} \in A) \sim N[\Phi_t x_t + \Lambda_t u_t, \Gamma_{t+1}] ,$$

and

(1.3.9) $$H_t(x_t,u_{t-1};z_t \in B) \sim N[\Psi_t x_t, R_t] \quad .$$

From (1.2.29) and (1.3.3) , G_0 satisfies

(1.3.10) $$G_0(x_0 \in A) \sim N(\hat{x}_0, \Pi_0) \quad .$$

The notation (1.3.8), for example, indicates that for x_t and u_t fixed, the stochastic kernel $Q_t(x_t,u_t;A)$ is the Gaussian measure in A with mean $\Phi_t x_t + \Lambda_t u_t$ and convariance Γ_{t+1} . That is

(1.3.11) $$Q_t(x_t,u_t;A) = \int_A \frac{1}{\sqrt{(2\pi)^n |\Gamma_{t+1}|}} [\exp - \tfrac{1}{2}(x_{t+1} - \Phi_t x_t - \Lambda_t u_t)' \cdot \Gamma_{t+1}^{-1}(x_{t+1} - \Phi_t x_t - \Lambda_t u_t)] dx_{t+1} \quad .$$

The integral is n-dimensional, and the prime indicates matrix transpose. The notation $N(\xi,\Sigma)$ will be used without any implication that Σ is non-singular . While there is no density in case $|\Sigma| = 0$, the distribution intended is clear.

Chapter 2 - Estimation

2.1 Introduction

In the next chapter, the distributions of x_t and of z_{t+1} both conditional on the data z^t are required for the development of the theory of sufficient statistics. The present chapter is devoted to computation of stochastic kernels $G_t(z^t,u^t;A)$, the filtering distribution, and $K_t(z^t,u^t;B)$ the control distribution, which represent these conditional distributions in the following sense:

$$P_{\{u\}}[x_t \in A|z^t] = G_t(z^t,u^{t-1}(z^{t-1});A) \qquad \text{a.s. } P^t_{\{u\}} \tag{2.1.1}$$

and

$$P_{\{u\}}[z_{t+1} \in B|z^t] = K_t(z^t,u^t(z^t);B) \qquad \text{a.s. } P^t_{\{u\}} \tag{2.1.2}$$

for $A \in \mathcal{B}_{\mathcal{X}_t}$, $B \in \mathcal{B}_{\mathcal{Z}_t}$, and all control laws $\{u\}$.

For the control problem these kernels are used principally to transform the incomplete information problem with loss function (1.2.32) into a non-Markovian dynamic programming problem with complete information, state variable z_t and transition distributions $K_t(z^t,u^{t-1};z_{t+1} \in B)$. This transformation will be developed systematically in Chapter 3 with the introduction of the notion of sufficient statistics. However, the method will be described here briefly and heuristically in order to motivate interest in the kernels G_t and K_t .

Clearly for all control laws $\{u\}$

$$E_{\{u\}}[\sum_{t=1}^{T} c_t(x_t,u^{t-1})] = E_{\{u\}}[\sum_{t=1}^{T} E_{\{u\}}[c_t(x_t,u^{t-1})|z^t]] \quad . \tag{2.1.3}$$

Thus the expected loss is preserved if the cost functions $c_t(x_t,u^{t-1})$ are replaced by conditional cost function (see Definition 5.2.1)

$$c^I_t(z^t,u^{t-1}) = \int c_t(x_t,u^{t-1})G_t(z^t,u^{t-1};dx_t) \tag{2.1.4}$$

where from (2.1.1) for all laws $\{u\}$

(2.1.5) $$E_{\{u\}}[c_t(x_t,u^{t-1})|z^t] = c_t^I(z^t,u^{t-1}(z^{t-1})) \qquad \text{a.s. } P_{\{u\}}^t .$$

The problem then is equivalent to the complete information with state variable z_t and loss function

(2.1.6) $$L^I = \sum_{t=1}^{T-1} c_t^I(z^t,u^{t-1}) .$$

Although this equivalence is not used formally, it is intended whenever dynamic programming results are compared with the incomplete information results of this work.

While the development of constructive means for computing the kernels K_t and G_t is presented here as a step in the solution of the optimal control problem, interest in the filtering distribution G_t can also be motivated quite independently of the control problem. Assume that a control law is given, that observations z^t have been taken, and that an estimate $\tilde{x}_t$ of the system process x_t is required so as to minimize an expected loss,

(2.1.7) $$E_{\{u\}}[L(x_t,\tilde{x}_t(z^t))] .$$

This is a standard Bayes point estimation problem. The optimum estimate is obtained by selecting the value of $\tilde{x}_t$ which minimized the posterior loss

(2.1.8) $$E_{\{u\}}[L(x_t,\tilde{x}_t)|z^t] = \int_{\mathcal{X}_t} L(x_t,\tilde{x}_t)P_{\{u\}}[dx_t|z^t] = \int_{\mathcal{X}_t} L(x_t,\tilde{x}_t)G_t(z^t,u^{t-1}(z^{t-1});dx_t)$$

for each point z^t (provided such a value exists and the resulting function $\tilde{x}_t(z^t)$ is measurable). Thus, for any loss function $L(x_t,\tilde{x}_t)$ the <u>posterior distribution</u> $P_{\{u\}}[x_t \in A|z^t]$ is required for the computation of the posterior loss (2.1.8) and from (2.1.1) the filtering distribution G_t determines the posterior distribution.

For example, for quadratic loss

(2.1.9) $$L(x_t,\tilde{x}_t) = |x_t - \tilde{x}_t|^2$$

the Bayes estimate is given by

$$\tilde{x}_t(z^t) = \int_{\mathcal{X}_t} x_t P_{\{u\}}[dx_t | z^t] = \int_{\mathcal{X}_t} x_t G_t(z^t, u^{t-1}(z^{t-1}); dx_t) \tag{2.1.10}$$

A discussion of the maximum likelihood criterion in extimation problems along with the smoothing problem in which the system x_t is estimated from the data z^T where $t < T$ is given by Rauch, Tung, and Striebel in [14] .

This chapter is concerned with a number of stochastic kernels $P_t(x_t, u_t; A \times B)$, $Q_t(x_t, u_t; A)$, $H_{t+1}(x_{t+1}, u_t; B)$, $J_t(z^t, u^t, A \times B)$, $G_t(z^t, u^t; A)$ and $K_t(z^t, u^t; B)$ all determined directly from the model (1.2.1)-(1.2.5). While they are functions of the control values u^t they are independent of the control law $\{u\}$. Thus, the class of admissible laws $\mathcal{D}$ and also the loss function L are irrelevant for this chapter. All kernels of interest have the further property that they only depend on the distribution of the processes x_τ and z_τ for $\tau = 0,,1,\ldots,t+1$. Thus the final time T is also irrelevant. As a convenience, it will be assumed throughout this chapter that time is unrestricted; that is, $t = 0,1,2,\ldots$.

Results of this chapter depend heavily on the properties of stochastic kernels which are collected for easy reference in section A.1 of the appendix. Section 2.2 is devoted to showing the existence of the filtering and control distributions and giving a constructive method for their computation for the general model (1.2.1)-(1.2.5) assuming only that the spaces $\mathcal{X}_t$ and Z_t are finite dimensional Euclidean spaces. This result is given by Dynkin [7] for countable spaces. In section 2.3, the results of section 2.2 are specialized under the assumption of densities for the kernels H_t and Q_t . Under the assumptions of the linear Gaussian model of section 1.3, results are further specialized in section 2.4 to obtain the Kalman filter equations. This result was first obtained by Kalman [9] using different methods and by Rauch, Tung, and Striebel [14] using essentially the methods of this chapter.

2.2 General filtering theory

In this section it will be shown by induction that if the system and observation processes x_t and z_t are vector valued, then filtering distributions G_t exist and satisfy (2.1.1). This is accomplished by giving a procedure by which G_{t+1} can be computed from G_t and P_t. The control distributions K_t satisfying (2.1.2) will also be produced by the procedure.

First, properties of P_t, Q_t and P_t will be developed.

Lemma 2.2.1 $P_t(x_t,u_t;C)$, $Q_t(x_t,u_t;A)$ and $H_{t+1}(x_{t+1},u_t;B)$ defined by (1.2.23)-(1.2.25) are stochastic kernels on $(\mathcal{X}_t \times \mathcal{U}_t, \mathcal{X}_{t+1} \times Z_{t+1})$, $(\mathcal{X}_t \times \mathcal{U}_t, \mathcal{X}_{t+1})$, and $(\mathcal{X}_{t+1} \times \mathcal{U}_t, Z_{t+1})$ respectively which satisfy

(2.2.1) $$P_t(x_t,u_t;A \times B) = \int_A H_{t+1}(x_{t+1},u_t;B)Q_t(x_t,u_t;dx_{t+1})$$

for $A \in \mathcal{B}_{\mathcal{X}_{t+1}}$, $B \in \mathcal{B}_{Z_{t+1}}$.

Proof: Since the functions φ_t and ψ_{t+1} are measurable, it follows from Lemma A.1.4 that Q_t and H_{t+1} given by (1.2.24) and (1.2.25) are stochastic kernels. Let $P_t(x_t,u_t;A \times B)$ be defined by (2.2.1). From Lemma A.1.3 P_t is a stochastic kernel. It will be shown that P_t satisfies (1.2.23). Throughout the remainder of the argument x_t and u_t are fixed, $w' = w_{t+1}$ and $w'' = e_{t+1}$. First, Lemma A.1.8 will be applied with $x = x_{t+1}$, $X(w') = \varphi_t(x_t,u_t,w_{t+1})$ and $\psi(x,w'') = \psi_{t+1}(x,u_t,e_{t+1})$. With this identification from (1.2.25)

(2.2.2) $$H(x,B) = H_{t+1}(x_{t+1},u_t;B) = P_{w''}[\{w''|\psi(x,w'') \in B\}] .$$

Thus from Lemma A.1.8

(2.2.3) $$H(x,B) = P_{\mathcal{W}}[Y \in B|X = x] \qquad \text{a.s. } P_{\mathcal{X}}$$

where

$$Y(w) = \psi(X(w'),w'') = \psi_{t+1}(\varphi_t(x_t,u_t,w_{t+1}),u_t,e_{t+1}) .$$

Next, Lemma A.1.7 will be applied with $X(w)$ degenerate, $Y(w) = \varphi_t(x_t,u_t,w_{t+1})$ and $Z(w) = \psi_{t+1}(\varphi_t(x_t,u_t,w_{t+1}),u_t,e_{t+1})$. For this identification (2.2.3) becomes (A.1.15) of Lemma A.1.7. From (1.2.24)

$$Q(A) = Q(x_t,u_t;A) = P_{\mathscr{W}}[\{w|Y(w) \in A\}] = P_{\mathscr{W}}[Y \in A|X=x] \quad \text{a.s. } P_{\mathfrak{X}} \tag{2.2.4}$$

Thus (2.2.1) becomes (A.1.3) , $J = P_t$, and from (A.1.16) of Lemma A.1.7 with $C = A \times B$

$$\begin{aligned} P_t(x_t,u_t;A \times B) &= J(x,A \times B) = P_{\mathscr{W}}[Y \in A,\ Z \in B|X = x] \\ &= P_{\Omega}[\{(w_{t+1},e_{t+1})|\varphi_t(x_t,u_t,w_{t+1}) \in A\ , \\ &\qquad \psi_{t+1}(\varphi_t(x_t,u_t,w_{t+1}),u_t,e_{t+1}) \in B\}] \quad \text{a.s. } P_{\mathfrak{X}} \ . \end{aligned} \tag{2.2.5}$$

Since $P_{\mathfrak{X}}$ is degenerate (2.2.5) holds identically, and it follows that P_t satisfies (1.2.23) .

<u>Lemma</u> 2.2.2 The stochastic kernels P_t , Q_t , H_{t+1} defined by (1.2.23)-(1.2.25) satisfy (1.2.26)-(1.2.28) .

Proof: Again Lemma A.1.8 will be applied - this time with $w" = (e^t,w^t)$, $w" = w_{t+1}$, and $x = (x_t,z^t)$. For a given control law $\{u\}$, from (1.2.13), (1.2.2), and (A.1.18)

$$(x_t,z^t) = X(w')$$

$$x_{t+1} = \varphi_t(x_t,u_t(z^t),w_{t+1}) = \psi(x,w") = Y(w)$$

$$Y(w) = \psi(X(w'),w") = \varphi_t(x_t,u_t(z^t),w_{t+1}) = x_{t+1}$$

With this identification, from (A.1.19) and (1.2.24)

$$\begin{aligned} H(x,B) &= P_{\mathscr{W}"}[\{(w"|\psi(x,w") \in B\}] = P_{\Omega}[\{w_{t+1}|\varphi(x_t,u_t(z^t),w_{t+1}) \in B\}] \\ &= Q_t(x_t,u_t(z^t);B) \ . \end{aligned} \tag{2.2.6}$$

Thus from (A.1.20) of Lemma A.1.8

$$H(x,B) = P_{\mathscr{W}}[Y \in B|X = x] = P_{\{u\}}[x_{t+1} \in B|x_t,z^t] \quad \text{a.s. } P_{\{u\}} \tag{2.2.7}$$

The result (1.2.27), then follows from (2.2.6) and (2.2.7) . Lemma A.1.8 will be applied again, this time with $w' = (e^t, w^{t+1})$, $w'' = e_{t+1}$, and $x = (x_{t+1}, x_t, z^t)$. From (1.2.13), (1.2.3), and (A.1.18)

$$(x_{t+1}, x_t, z^t) = X(w')$$

$$\psi_{t+1}(x_{t+1}, u_t(z^t), e_{t+1}) = \psi(x, w'')$$

$$Y(w) = \psi(X(w'), w'') = \psi_{t+1}(x_{t+1}, u_t(z^t), e_{t+1}) = z_{t+1}$$

With this identification, from (A.1.19) and (1.2.25)

$$H(x,B) = P_{\mathscr{W}}[\{w'' | \psi(x,w'') \in B\}] = P_{\Omega}[\{e_{t+1} | \psi_{t+1}(x_{t+1}, u_t(z^t), e_{t+1}) \in B\}]$$

$$= H_{t+1}(x_{t+1}, u_t(z^t); B) \quad .$$

Thus from Lemma A.1.8

$$H(x,B) = P_{\mathscr{W}}[Y \in B | X = x] = P_{\{u\}}[z_{t+1} \in B | x_{t+1}, x_t, z^t] ,$$

and (1.2.28) follows. The result (1.2.26) follows from A.1.7 with $x = (x_t, z^t)$, $y = x_{t+1}$, $z = x_{t+1}$,

$$H(x,y;B) = H_{t+1}(x_{t+1}, u_t(z^t); B) ,$$

and

$$Q(x,A) = Q_t(x_t u_t(z^t); A) .$$

Then from (2.2.1) of Lemma 2.2.1 , $J(x,C)$ defined by (A.1.3) satisfies

$$J(x, A \times B) = \int_A H(x,y;B)Q(x,dy) = \int_A H_{t+1}(x_{t+1}, u_t(z^t); B) Q_t(x_t, u_t(z^t); dx_{t+1})$$

$$= P_t(x_t, u_t(z^t); A \times B) ,$$

and from (A.1.16) of Lemma A.1.7

$$J(x, A \times B) = P_{\{u\}}[x_{t+1} \in A , z_{t+1} \in B | x_t, z^t] \qquad \text{a.s. } P_{\{u\}}$$

Conditions (A.1.14) and (A.1.15) of Lemma A.1.7 are supplied by (1.2.27) and (1.2.28) , respectively.

The next result which follows from lemmas of the appendix, is required both in the derivation of the filtering distributions G_t and later in the treatment of sufficient statistics.

Lemma 2.2.3 Let $\mathcal{X}_{t+1}$ and Z_{t+1} be finite dimensional Euclidean vector spaces, let $G_t^*(y_t,A)$ be a stochastic kernel on $(\mathcal{Y}_t,\mathcal{X}_t)$ where $(\mathcal{Y}_t,\mathcal{B}_{\mathcal{Y}_t})$ is a measurable space, let P_t be a stochastic kernel $(\mathcal{X}_t \times \mathcal{U}_t;\mathcal{X}_{t+1} \times Z_{t+1})$, and define J_t^* by

$$(2.2.8) \qquad J_t^*(y_t,u_t;A \times B) = \int_{\mathcal{X}_t} P_t(x_t,u_t;A \times B) G_t^*(y_t,dx_t)$$

Then J_t^* is a stochastic kernel on $(\mathcal{Y}_t \times \mathcal{U}_t, \mathcal{X}_{t+1} \times Z_{t+1})$, and there exists $G_{t+1}^*(y_t,z_{t+1},u_t;A)$ a stochastic kernel on $(\mathcal{Y}_t \times Z_{t+1} \times \mathcal{U}_t,\mathcal{X}_{t+1})$ which satisfies

$$(2.2.9) \qquad G_{t+1}^*(y_t,z_{t+1},u_t;A) = \frac{J_{t,A}^*(y_t,u_t;dz_{t+1})}{K_t^*(y_t,u_t;dz_{t+1})}(z_{t+1}) \quad \text{a.s. } K_t^*(y_t,u_t;B)$$

for all (y_t,u_t) and $A \in \mathcal{B}_{\mathcal{X}_t}$ where

$$(2.2.10) \qquad J_{t,A}^*(y_t,u_t;B) = J_t^*(y_t,u_t;A \times B)$$

and

$$(2.2.11) \qquad K_t^*(y_t,u_t;B) = J_t^*(y_t,u_t;\mathcal{X}_{t+1} \times B)$$

[The notation a.s. K_t^* in (2.2.9) indicates that for each (y_t,u_t) the set N of values of z_{t+1} for which (2.2.9) fails to hold satisfies $K_t^*(y_t,u_t;N) = 0$. The right side of (2.2.9) indicates the Radon-Nikodym derivative of $J_{t,A}^*(y_t,u_t;B)$ with respect to $K_t^*(y_t,u_t;B)$ for y_t,u_t and A fixed.]

Proof: From Lemma A.1.2 ii) J_t^* is a stochastic kernel. The existence of G_{t+1}^*, a stochastic kernel satisfying (2.2.9), follows from Lemma A.1.9 since by assumption $\mathcal{Y} = Z_{t+1}$, $Z = \mathcal{X}_{t+1}$ are finite dimensional Euclidean spaces and $\mathcal{X} = \mathcal{Y}_t \times \mathcal{U}_t$ is a measurable space.

Attention is turned now to derivation of a recursive equation for the filtering distribution.

Lemma 2.2.4 Let $G_t(z^t,u^{t-1};A)$ be a stochastic kernel on $(\mathcal{Z}^t \times \mathcal{U}^{t-1}),\mathcal{X}_t)$ which satisfies (2.1.1), and let P_t, Q_t and H_t be defined by (1.2.23)-(1.2.25). Then

$$(2.2.12) \qquad J_t(z^t,u^t;A \times B) = \int_{\mathcal{X}_t} P_t(x_t,u_t;A \times B)G_t(z^t,u^{t-1};dx_t)$$

$$= \int_{\mathcal{X}_t} [\int_A H_{t+1}(x_{t+1},u_t;B)Q_t(x_t,u_t;dx_{t+1})]G_t(z^t,u^{t-1};dx_t)$$

and

$$(2.2.13) \qquad K_t(z^t,u^t;B) = J_t(z^t,u^t;\mathcal{X}_{t+1} \times B)$$

$$= \int_{\mathcal{X}_t} P_t(x_t,u_t;\mathcal{X}_{t+1} \times B)G_t(z^t,u^{t-1};dx_t)$$

$$= \int_{\mathcal{X}_t} [\int_{\mathcal{X}_{t+1}} H_{t+1}(x_{t+1},u_t;B)Q_t(x_t,u_t;dx_{t+1})]G_t(z^t,u^{t-1};dx_t)$$

are stochastic kernels on $(\mathcal{Z}^t \times \mathcal{U}^t, \mathcal{X}_{t+1} \times \mathcal{Z}_{t+1})$ and $(\mathcal{Z}^t \times \mathcal{U}^t,\mathcal{Z}_{t+1})$ respectively. The kernel K_t satisfies (2.1.2) and J_t satisfies

$$(2.2.14) \qquad P_{\{u\}}[x_{t+1} \in A,\ z_{t+1} \in B|z^t] = J_t(z^t,u^t(z^t);A \times B) \quad \text{a.s. } P_{\{u\}} .$$

Proof: From Lemma A.1.2 ii) J_t is a stochastic kernel. The second line of (2.2.12) follows from (2.2.1) of Lemma 2.2.1.

For a given control law $\{u\}$ (2.2.14) will be verified by applying Lemma A.1.7 with $x = z^t$, $y = x_t$, $z = (x_{t+1},z_{t+1})$, $Q(x,A) = G_t(z^t,u^{t-1}(z^{t-1});A)$, $H(x,y;B) = P_t(x_t,u_t(z^t);B)$, $\mathcal{W} = \mathcal{X}^T \times \mathcal{Z}^T \times \mathcal{U}^{T-1}$, $P_{\mathcal{W}} = P_{\{u\}}$. The conditions (A.1.14) and (A.1.15) are supplied by the assumption (2.1.1) and by (1.2.26) which follows from Lemma 2.2.2. From (A.1.3) and (2.2.13)

$$(2.2.15) \qquad J(x,\mathcal{Y} \times (A \times B)) = \int_{\mathcal{Y}} H(x,y;A \times B)Q(x,dy)$$

$$= \int_{\mathcal{X}_t} P_t(x_t,u_t(z^t);A \times B)G_t(z^t,u^{t-1}(z^{t-1});dx_t) = J_t(z^t,u^t(z^t);A \times B)$$

and from (A.1.16) of Lemma A.1.7

$$J(x,\mathcal{Y} \times (A \times B)) = P_{\mathcal{W}}[Z \in A \times B | X = x]$$

(2.2.16)

$$= P_{\{u\}}[x_{t+1} \in A, z_{t+1} \in B | z^t] \qquad \text{a.s. } P_{\{u\}} .$$

The result (2.2.14) then follow from (2.2.15) and (2.2.16).

Equations (2.2.13) and (2.1.2) follow from (2.2.12) and (2.2.14) respectively.

Lemma 2.2.5 Let $J_t(z^t,u^t;A \times B)$ be a stochastic kernel which satisfies (2.2.14) and let G_{t+1} be a stochastic kernel which satisfies

(2.2.17) $$G_{t+1}(z^{t+1},u^t;A) = \frac{J_{t,A}(z^t,u^t;dz_{t+1})}{K_t(z^t,u^t;dz_{t+1})}(z_{t+1}) \qquad \text{a.s. } K_t(z^t,u^t;B)$$

where

(2.2.18) $$J_{t,A}(z^t,u^t;B) = J_t(z^t,u^t;A \times B)$$

(2.2.19) $$K_t(z^t,u^t;B) = J_t(z^t,u^t;\mathcal{X}_{t+1} \times B) \quad .$$

Then G_{t+1} satisfies (2.1.1) for $t+1$.

Proof: Since $G_{t+1}(z^{t+1},u^t;A)$ is measurable in (z^{t+1},u^t) for each A and $u^t(z^t)$ is measurable, $G_{t+1}(z^{t+1},u^t(z^t);A)$ is measurable in z^{t+1} .

Lemma A.1.6 will be applied with $x = z^t$, $y = (x_{t+1},z_{t+1})$, $Q(x,C) = J_t(z^t,u^t(z^t);C)$ and $P_{\mathcal{W}} = P_{\{u\}}$. With this idenfication (2.2.14) becomes (A.1.9). Then from (A.1.11) of Lemma A.1.6 for $A \in \mathcal{B}_{\mathcal{X}_t}$, $B^t \in \mathcal{B}_Z^t$, and $B \in \mathcal{B}_{Z_{t+1}}$

$$P_{\{u\}}[x_{t+1} \in A,\ z^t \in B^t,\ z_{t+1} \in B]$$

(2.2.20)

$$= \int_{B_t} J_t(z^t,u^t(z^t);A \times B)P_{\{u\}}[dz^t] \ .$$

For $A = \mathcal{X}_{t+1}$, from (2.2.19) this becomes

$$(2.2.21) \qquad P_{\{u\}}[z^t \in B^t, z_{t+1} \in B] = \int_{B^t} K_t(z^t, u^t(z^t); B) P_{\{u\}}[dz^t] .$$

Thus from (2.2.21), (A.1.12) of Lemma A.1.6, (2.2.17) , (2.2.18), and (2.2.20)

$$\int_{B^t} \int_B G_{t+1}(z^t, z_{t+1}, u^t(z^t); A) P_{\{u\}}[dz^t, dz_{t+1}]$$

(2.2.22)

$$= \int_{B^t} [\int_B G_{t+1}(z^t, z_{t+1}, u^t(z^t); A) K_t(z^t, u^t(z^t); dz_{t+1})] P_{\{u\}}[dz^t]$$

$$= \int_{B^t} J_{t,A}(z^t, u^t(z^t); B) P_{\{u\}}[dz^t]$$

$$= P_{\{u\}}[x_{t+1} \in A, \ z^{t+1} \in B^t \times B] \ .$$

Since (2.2.22) holds for all product sets $B^t \times B \in \mathcal{B}_Z^{t+1} = \mathcal{B}_Z^t \times \mathcal{B}_{Z_{t+1}}$ it follows that

$$\int_{B^{t+1}} G_{t+1}(z^{t+1}, u^t(z^t); A) P_{\{u\}}[dz^{t+1}] = P_{\{u\}}[x_{t+1} \in A, z^{t+1} \in B^{t+1}]$$

for all $B^{t+1} \in \mathcal{B}_Z^{t+1}$. Thus (2.1.1) frollows from the definition of conditional probability.

<u>Theorem</u> 2.2.1 Let $\mathcal{X}_0, \mathcal{X}_1, \ldots, Z_1, Z_2, \ldots$ be finite dimensional Euclidean spaces, let G_0 be a probability on $(\mathcal{X}_0, \mathcal{B}_{\mathcal{X}_0})$, and let P_t be stochastic kernels on $(\mathcal{X}_t \times \mathcal{U}_t, \mathcal{X}_{t+1} \times Z_{t+1})$ for $t = 0,1,2,\ldots$. Then there exist stochastic kernels J_t , K_t , and G_t which satisfy (2.2.12), (2.2.13), and (2.2.17) - (2.2.19) for $t = 0,1,2,\ldots$.

Proof: The proof is by induction. By assumption G_0 is a stochastic kernel. Assume then that G_t is a stochastic kernel. From Lemma 2.2.3 with $y_t = (z^t, u^{t-1})$ there exists kernels J_t , K_t , and G_{t+1} which satisfy (2.2.12), (2.2.13), (2.2.17) and (2.2.18) .

<u>Theorem</u> 2.2.2 Let P_t , Q_t , H_t , and G_0 satisfy (1.2.23)-(1.2.25) and (1.2.29) and suppose there exist stochastic kernels G_t , K_t and J_t that satisfy (2.2.12), (2.2.13) and (2.2.17)-(2.2.19) for $t = 0,1,2,\ldots$. Then

G_t, K_t and J_t satisfy (2.1.1),(2.1.2), and (2.2.14) for $t = 0,1,2,\ldots$.

Proof: The proof is again by induction. From (1.2.29) G_0 satisfies (2.1.1) for $t = 0$. Assume then that G_t satisfies (2.1.1). From Lemma 2.2.4 J_t and K_t satisfy (2.2.14) and (2.1.2), and from Lemma 2.2.5 G_{t+1} satisfies (2.1.1) for $t+1$.

Theorem 2.2.3 Let $\mathcal{X}_0,\mathcal{X}_1,\ldots,\mathcal{Z}_1,\mathcal{Z}_2,\ldots$ be finite dimensional Euclidean spaces and let P_t, Q_t, H_t and G_0 satisfy (1.2.23)-(1.2.25) and (1.2.29) for $t=0,1,\ldots$. Then there exist stochastic kernels G_t and K_t which satisfy (2.1.1) and (2.1.2) for $t = 0,1,2,\ldots$. These kernels can be computed recursively from (2.2.12), (2.2.13), and (2.2.17)-(2.2.19) .

Proof: This follows from Theorems 2.2.1 and 2.2.2 .

2.3 Estimation formulas

If densities can be found for the stochastic kernels Q_t and H_t , considerable simplification of the procedure for computing the distributions G_t and K_t results. First, it will be assumed only that the kernels H_t have densities.

Lemma 2.3.1. Let $G_t(z^t,u^{t-1};A)$, $Q_t(x_t,u_t;A)$ and $H_{t+1}(x_{t+1},u_t;B)$ be stochastic kernels, let $h_{t+1}(x_{t+1},u_t;z_{t+1})$ be a non-negative, jointly measurable function, and let λ_{t+1} be a σ-finite measure on $(\mathcal{Z}_{t+1},\mathcal{B}_{\mathcal{Z}_{t+1}})$ which satisfy

$$(2.3.1)\qquad H_{t+1}(x_{t+1},u_t;B) = \int_B h_{t+1}(x_{t+1},u_t;z_{t+1})\lambda_{t+1}(dz_{t+1}) ,$$

for all $B \in \mathcal{B}_{\mathcal{Z}_{t+1}}$. Then for $A \in \mathcal{B}_{\mathcal{X}_{t+1}}$

$$(2.3.2)\qquad j_t^*(z^t,u^t;A,z_{t+1}) = \int_{\mathcal{X}_t}\Big[\int_A h_{t+1}(x_{t+1},u_t;z_{t+1})Q_t(x_t,u_t;dx_{t+1})\Big] \cdot G_t(z^t,u^{t-1};dx_t)$$

is jointly measurable in (z^t,u^t,z_{t+1}) for A fixed, is a measure in A for (z^{t+1},u^t) fixed, and satisfies

(2.3.3) $$J_t(z^t,u^t;A \times B) = \int_B j_t^*(z^t,u^t;A,z_{t+1})\lambda_{t+1}(dz_{t+1})$$

where J_t is defined by (2.2.12).

Proof: Since

$$\chi_A(x_{t+1})h_{t+1}(x_{t+1},u_t;z_{t+1})$$

is jointly measurable in (x_{t+1},u_t,z_{t+1}) for all A and a measure in A for all (x_{t+1},u_t,z_{t+1}), from Lemma A.1.2 i)

$$[\int_A h_{t+1}(x_{t+1},u_t;z_{t+1})Q_t(x_t,u_t;dx_{t+1})]$$

is a measure in A for all (x_t,u_t,z_{t+1}) and is measurable in (x_t,u_t,z_{t+1}) for all A. From the same lemmas $j_t^*(z^t,u^t;A,z_{t+1})$ given by (2.3.2) is a measure in A and measurable in (z^{t+1},u^t).

From (2.3.1) and the Fubini Theorem with respect to the product measure $Q_t \times \lambda_{t+1}$ for (x_t,u_t) fixed

$$\int_B [\int_A h_{t+1}(x_{t+1},u_t;z_{t+1})Q_t(x_t,u_t;dx_{t+1}]\lambda_{t+1}(dz_{t+1})$$

(2.3.4) $$= \int_A [\int_B h_{t+1}(x_{t+1},u_t;z_{t+1})\lambda_{t+1}(dz_{t+1})]Q_t(x_t,u_t;dx_{t+1})$$

$$= \int_A H_{t+1}(x_{t+1},u_t;B)Q_t(x_t,u_t;dx_{t+1}) \quad .$$

For $B \in \mathcal{B}_{Z_{t+1}}$, from (2.3.2), the Fubini Theorem for $G_t \times \lambda_{t+1}$ and (z^t,u^{t-1}) fixed, (2.3.4), and the definition (2.2.12)

$$\int_B j_t^*(z^t,u^t;A,z_{t+1})\lambda_{t+1}(dz_{t+1})$$

$$= \int_B \{\int_{\mathcal{X}_t} [\int_A h_{t+1}(x_{t+1},u_t;z_{t+1})Q_t(x_t,u_t;dx_{t+1})]G_t(z^t,u^{t-1};dx_t)]$$

$$\cdot \lambda_{t+1}(dz_{t+1})$$

$$= \int_{\mathcal{X}_t} \{\int_B [\int_A h_{t+1}(x_{t+1},u_t;z_{t+1})Q_t(x_t,u_t;dx_{t+1}]\lambda_{t+1}(dz_{t+1})\}$$

$$\cdot G_t(z^t,u^{t-1};dx_t)$$

$$= \int_{\mathcal{X}_t} \{\int_A H_{t+1}(x_{t+1},u_t;B)Q_t(x_t,u_t;dx_{t+1}\}G_t(z^t,u^{t-1};dx_t) = J_t(z^t,u^t,A \times B)$$

Theorem 2.3.1. Let G_0 be a probability on $(\mathcal{X}_0, \mathcal{B}_{\mathcal{X}_0})$, let H_t and Q_t be stochastic kernels, and let h_t be a non-negative, jointly measurable function which satisfies (2.3.1). The for j_t^* given by (2.3.2), stochastic kernels G_t can be computed which satisfy

$$G_{t+1}(z^{t+1},u^t;A) = \frac{j_t^*(z^t,u^t; A, z_{t+1})}{j_t^*(z^t,u^t;\mathcal{X}_{t+1},z_{t+1})}$$

(2.3.5)

$$= \frac{\int_{\mathcal{X}_t}[\int_A h_{t+1}(x_{t+1},u_t;z_{t+1})Q_t(x_t,u_t;dx_{t+1})]G_t(z^t,u^{t-1};dx_t)}{\int_{\mathcal{X}_t}[\int_{\mathcal{X}_{t+1}} h_{t+1}(x_{t+1},u_t;z_{t+1})Q_t(x_t,u_t;dx_{t+1})]G_t(z^t,u^{t-1};dx_t)}$$

$$t = 0,1,2,\ldots$$

Further, if Q_t , H_t and G_0 are given by (1.2.24), (1.2.25), and (1.2.29), then G_t defined by (2.3.5) satisfies (2.1.1) and the stochastic kernel K_t defined by

(2.3.6) $$K_t(z^t,u^t;B) = \int_B j_t^*(z^t,u^t;\mathcal{X}_{t+1},z_{t+1})\lambda(dz_{t+1})$$

satisfies (2.1.2) for all control laws $\{u\}$. [It is understood that for values of (z^{t+1},u^t) for which the denominator of (2.3.5) is zero or infinite, $G_{t+1}(z^{t+1},u^t;\cdot)$ is given the value of an arbitrary probability measure.]

Proof: It will be shown first by induction that there exist stochastic kernels G_t which satisfy (2.3.5) and (2.2.17)-(2.2.19) where J_t is defined by (2.3.3). By assumption G_0 is a stochastic kernel. From Lemma 2.3.1, j_0^* and J_0 defined by (2.3.2) and (2.3.3) satisfy the assumptions of Lemma A.1.10. Thus from Lemma A.1.10 G_1 defined by (2.3.5) is a stochastic kernel which satisfies (2.2.17)-(2.2.19) where J_t is given by (2.3.3). Assume as an induction hypothesis that G_t is a stochastic kernel. It follows by the same argument that G_{t+1} defined by (2.3.2) and (2.3.5) is a stochastic kernel and satisfies (2.2.17)-(2.2.19). From Lemma 2.3.1 , J_t given by (2.3.3) satisfies (2.2.12), and from (2.3.3)

K_t defined by (2.3.6) satisfies (2.2.13). Thus it has been established that G_t, J_t and K_t defined in this way satisfy (2.2.12), (2.2.13), (2.2.17), and (2.2.18). The result follows then from Theorem 2.2.2.

If the distributions Q_t and G_0 also have densities, then further simplifications of the formula (2.3.5) are possible.

Lemma 2.3.2. Let $h_{t+1}(x_{t+1},u_t;z_{t+1}), q_t(x_t,u_t;x_{t+1}), g_t(z^t,u^{t-1};x_t)$ be non-negative, jointly measurable functions which satisfy (2.3.1),

(2.3.7) $$Q_t(x_t,u_t;A) = \int_A q_t(x_t,u_t;x_{t+1})\mu_{t+1}(dx_{t+1}) ,$$

and

(2.3.8) $$G_t(z^t,u^{t-1};A) = \int_B g_t(z^t,u^{t-1};x_t)\mu_t(dx_t)$$

for $A \in \mathcal{B}_{\mathcal{X}_{t+1}}$ and $B \in \mathcal{B}_{\mathcal{X}_t}$ where λ_{t+1}, u_t and μ_{t+1} are σ-finite measures, and H_t, Q_t, and G_t are stochastic kernels. Then

(2.3.9) $$j_t(z^t,u^t;x_{t+1},z_{t+1}) = h_{t+1}(x_{t+1},u_t;z_{t+1})\int_{\mathcal{X}_t} q_t(x_t,u_t;x_{t+1}) \cdot g_t(z^t,u^{t-1};x_t)\mu_t(dx_t)$$

is non-negative, measurable, and satisfies

(2.3.10) $$J_t(z^t,u^t;A\times B) = \iint_{AB} j_t(z^t,u^t;x_{t+1},z_{t+1})\lambda_{t+1}(dz_{t+1})\mu_{t+1}(dx_{t+1}) = \int_B j_t^*(z^t,u^t;A,z_{t+1})\lambda_{t+1}(dz_{t+1})$$

and

(2.3.11) $$j_t^*(z^t,u^t;A,z_{t+1}) = \int_A j_t(z^t,u^t;x_{t+1},z_{t+1})\mu_{t+1}(dx_{t+1})$$

where J_t and j_t^* are defined by (2.2.12) and (2.3.2). Further,

(2.3.12) $$\begin{aligned} k_t(z^t,u^t;z_{t+1}) &= \int_{\mathcal{X}_{t+1}} j_t(z^t,u^t;x_{t+1},z_{t+1})\mu_{t+1}(dx_{t+1}) \\ &= \int_{\mathcal{X}_{t+1}}\int_{\mathcal{X}_t} h_{t+1}(x_{t+1},u_t;z_{t+1})q_t(x_t,u_t;x_{t+1})g_t(z^t,u^{t-1};x_t)\mu_t(dx_t)\mu_{t+1}(dx_{t+1}) \\ &= j_t^*(z^t,u^t;\mathcal{X}_{t+1},z_{t+1}) \end{aligned}$$

is jointly measurable and satisfies

(2.3.13) $$K_t(z^t,u^t;B) = \int_B k_t(z^t,u^t;z_{t+1})\lambda_{t+1}(dz_{t+1})$$

where K_t is defined by (2.2.13).

Proof: From the Fubini Theorem and the measurability of the non-negative functions h_{t+1}, q_t and g_t, j_t given by (2.3.9) is measurable and non-negative. From an elementary property of the density and (2.3.8),

(2.3.14) $$\int_{\mathcal{X}_t} q_t(x_t,u_t;x_{t+1})g_t(z^t,u^{t-1};x_t)\mu_t(dx_t)$$
$$= \int_{\mathcal{X}_t} q_t(x_t,u_t;x_{t+1})G_t(z^t,u^{t-1};dx_t) \quad .$$

Similarly, from (2.3.7)

(2.3.15) $$\int_A h_{t+1}(x_{t+1},u_t;z_{t+1})q_t(x_t,u_t;x_{t+1})\mu_{t+1}(dx_{t+1})$$
$$= \int_A h_{t+1}(x_{t+1},u_t;z_{t+1})Q_t(x_t,u_t;dx_{t+1}) \quad .$$

Thus, from (2.3.9), (2.3.14), the Fubini Theorem for $G_t \times \mu_{t+1}$ and (z^t,u^{t-1}) fixed, (2.3.15), and (2.3.2)

$$\int_A j_t(z^t,u^t;x_{t+1},z_{t+1})\mu_{t+1}(dx_{t+1})$$
$$= \int_A h_{t+1}(x_{t+1},u_t;z_{t+1})[\int_{\mathcal{X}_t} q_t(x_t,u_t;x_{t+1})g_t(z^t,u^{t-1};x_t)\mu_t(dx_t)]\mu_{t+1}(dx_{t+1})$$
$$= \int_A h_{t+1}(x_{t+1},u_t;z_{t+1})[\int_{\mathcal{X}_t} q_t(x_t,u_t;x_{t+1})G_t(z^t,u^{t-1};dx_t)]\mu_{t+1}(dx_{t+1})$$
$$= \int_{\mathcal{X}_t}[\int_A h_{t+1}(x_{t+1},u_t;z_{t+1})q_t(x_t,u_t;x_{t+1})\mu_{t+1}(dx_{t+1})]G_t(z^t,u^{t-1};dx_t)$$
$$= \int_{\mathcal{X}_t}[\int_A h_{t+1}(x_{t+1},u_t;z_{t+1})Q_t(x_t,u_t;dx_{t+1}]G_t(z^t,u^{t-1};dx_t)$$
$$= j_t^*(z^t,u^t;A,z_{t+1}) \quad .$$

Thus (2.3.11) holds. The result (2.3.10) follows from (2.3.11), (2.3.3) of Lemma 2.3.1, and the Fubini Theorem for $\lambda_{t+1} \times \mu_{t+1}$ and (z^t, u^t) fixed. The result (2.3.13) follows from (2.3.12), (2.3.10) and (2.2.13) .

<u>Theorem</u> 2.3.2. Let Q_t , H_t and G_0 given by (1.2.24), (1.2.25), and (1.2.29) satisfy (2.3.7), (2.3.1), and

$$(2.3.16) \qquad \int G_0(A) = \int_A g_0(x_0)\mu_0(dx_0) \qquad A \in \mathcal{B}_{\mathcal{X}_0}$$

where h_t , q_t , and g_0 are non-negative, measurable functions, and μ_t , λ_t , and μ_0 are σ-finite measures for $t = 1,2,\ldots$. Then there exist $g_t(z^t, u^{t-1}; x_t)$ non-negative and measurable which satisfy

$$g_{t+1}(z^{t+1}, u^t; x_{t+1}) = \frac{j_t(z^t, u^t; x_{t+1}, z_{t+1})}{\int_{\mathcal{X}_{t+1}} j_t(z^t, u^t; x_{t+1}, z_{t+1})\mu_{t+1}(dx_{t+1})}$$

(2.3.17)

$$= \frac{h_{t+1}(x_{t+1}, u_t; z_{t+1})\int_{\mathcal{X}_t} q_t(x_t, u_t; x_{t+1}) g_t(z^t, u^{t-1}; x_t)\mu_t(dx_t)}{\int_{\mathcal{X}_{t+1}}\int_{\mathcal{X}_t} h_{t+1}(x_{t+1}, u_t; z_{t+1}) q_t(x_t, u_t; x_{t+1}) g_t(z^t, u^{t-1}; x_t)\mu_t(dx_t)\mu_{t+1}(dx_{t+1})}$$

where the j_t are given by (2.3.9). Further G_t defined by (2.3.8) and K_t defined by (2.3.12) and (2.3.13) satisfy (2.1.1) and (2.1.2) for all control laws $\{u\}$.

[It is understood that for values of (z^{t+1}, u^t) for which the denominator of (2.3.17) is zero or infinite g_{t+1} is given the value of an arbitrary probability density with respect to μ_{t+1}.]

Proof: It will be shown by induction that there exist g_t non-negative and measurable for which (2.3.17) holds, and G_t given by (2.3.8) is a stochastic kernel which satisfies (2.3.5) of Theorem 2.3.1. By assumption G_0 is a stochastic kernel. Assume as an induction hypothesis that g_t is a non-negative measurable function and that G_t given by (2.3.8) is a stochastic kernel. Then from (2.3.11) of Lemma 2.3.2 and the measurability of j_t^*

$$\{(z^t,u^t,x_{t+1}) | \int_{\mathcal{X}_{t+1}} j_t(z^t,u^t;x_{t+1},z_{t+1})\mu_{t+1}(dx_{t+1})$$

$$= j_t^*(z^t,u^t;\mathcal{X}_{t+1},z_{t+1}) = \infty \text{ or } 0\}$$

is measurable. It follows that g_{t+1} defined by (2.3.17) is measurable. Also from Lemma 2.3.2 j_t is non-negative so that g_{t+1} is non-negative. For (z^{t+1},u^t) such that

$$0 < j_t^*(z^t,u^t;\mathcal{X}_{t+1},z_{t+1}) < \infty \tag{2.3.18}$$

and G_{t+1} given by (2.3.8), from (2.3.17) and (2.3.11) of Lemma 2.3.2

$$\begin{aligned} G_{t+1}(z^{t+1},u^t;A) &= \int_A g_{t+1}(z^{t+1},u^t;x_{t+1})\mu_{t+1}(dx_{t+1}) \\ &= \int_A \left[\frac{j_t(z^t,u^t;x_{t+1},z_{t+1})}{j_t^*(z^t,u^t;\mathcal{X}_{t+1},z_{t+1})}\right]\mu_{t+1}(dx_{t+1}) \\ &= \frac{\int_A j_t(z^t,u^t;x_{t+1},z_{t+1})\mu_{t+1}(dx_{t+1})}{j_t^*(z^t,u^t;\mathcal{X}_{t+1},z_{t+1})} \\ &= \frac{j_t^*(z^t,u^t;A,z_{t+1})}{j_t^*(z^t,u^t;\mathcal{X}_{t+1},z_{t+1})} \quad . \end{aligned}$$

Thus G_{t+1} satisfies (2.3.5). For values of (z^{t+1},u^t) for which (2.3.18) does hot hold $g_{t+1}(z^{t+1},u^t;x_{t+1})$ is a density in x_{t+1}. Thus $G_{t+1}(z^{t+1},u^t;A)$ is a probability measure in A and satisfies (2.3.5) for these values. From (2.3.12) and (2.3.13) K_t satisfies (2.3.6). The result then follows from Theorem 2.3.1.

2.4 Estimation for the linear Gaussian model: the Kalman filter

It will be assumed now that the conditions (1.3.1)-(1.3.5) for a linear Gaussian model hold where the matrices Π_0 , R_t , and Γ_t are non-singular. It follows then that the stochastic kernels Q_t and H_t satisfy (1.3.8) and (1.3.9). Thus the assumptions (2.3.1) and (2.3.7) of Theorem 2.3.2 are satisfied where

$$(2.4.1)\qquad h_t(x_t,z_t) = \frac{1}{(2\pi)^{m/2}\sqrt{|R_t|}} \exp -\tfrac{1}{2}(z_t-\Psi_t x_t)' R_t^{-1}(z_t-\Psi_t x_t) \quad t = 1,2,\ldots ,$$

$$(2.4.2)\qquad q_t(x_t,u_t;x_{t+1}) = \frac{1}{(2\pi)^{n/2}\sqrt{|\Gamma_{t+1}|}} \exp -\tfrac{1}{2}(x_{t+1}-\Phi_t x_t-\Lambda_t u_t)'$$

$$\cdot\ \Gamma_{t+1}^{-1}(x_{t+1} - \Phi_t x_t-\Lambda_t u_t) \quad t = 0,1,2,\ldots,$$

and the measures λ_t and μ_{t+1} are Lebesgue measures of m and n dimensions respectively. From (1.3.10),(2.3.16) of Theorem 2.3.2 is satisfied where

$$(2.4.3)\qquad g_0(x_0) = \frac{1}{(2\pi)^{n/2}\sqrt{|\Pi_0|}} \exp -\tfrac{1}{2}(x_0-\hat{x}_0)'\Pi_0^{-1}(x_0-\hat{x}_0)$$

and μ_0 is Lebesgue measure of dimension n . Thus Theorem 2.3.2 applies.

<u>Theorem</u> 2.4.1. For the linear Gaussian model (1.3.1)-(1.3.5) where it is assumed that the covariance matrices Π_0 , R_t, Γ_t are nonsingular for $t = 1,2,\ldots$, there exist stochastic kernels G_t and K_t which satisfy (2.1.1) and (2.1.2) for all control laws $\{u\}$. The filtering distributions G_t are given by

$$(2.4.4)\qquad G_t(z^t,u^{t-1};x_t \in A) \sim N(\hat{x}_t(z^t,u^{t-1}),\Pi_t) \quad t = 1,2,3,\ldots$$

where $\hat{x}_t(z^t,u^{t-1})$, and Π_t satisfy

$$(2.4.5)\qquad \hat{x}_{t+1} = \Phi_t\hat{x}_t + \Lambda_t u_t + \Pi_{t+1}\Psi'_{t+1}R_{t+1}^{-1}(z_{t+1}-\Psi_{t+1}\Lambda_t u_t-\Psi_{t+1}\Phi_t\hat{x}_t) \quad t = 0,1,2,\ldots,$$

and

$$(2.4.6)\qquad \Pi_{t+1}^{-1} = \Psi'_{t+1}R_{t+1}^{-1}\Psi_{t+1} + (\Gamma_{t+1} + \Phi_t\Pi_t\Phi'_t)^{-1} \qquad t = 0,1,2,\ldots$$

The initial values $\hat{x}_0$ and Π_0 are supplied by assumption (1.3.3).

Further, J_t defined by (2.2.13) and the control distributions K_t satisfy

$$(2.4.7)\qquad J_t(z^t,u^t;x_{t+1}\in A,z_{t+1}\in B) \sim N\left[\begin{pmatrix}\Lambda_t u_t + \Phi_t\hat{x}_t \\ \Psi_{t+1}(\Lambda_t u_t + \Phi_t\hat{x}_t)\end{pmatrix}, \ddagger_{t+1}\right]$$

where

$$(2.4.8)\qquad \ddagger_{t+1} = \begin{bmatrix}(\Gamma_{t+1}+ \Phi_t\Pi_t\Phi'_t) & (\Gamma_{t+1}+\Phi_t\Pi_t\Phi'_t)\Psi'_{t+1} \\ \Psi_{t+1}(\Gamma_{t+1}+\Phi_t\Pi_t\Phi'_t) & R_{t+1} + \Psi_{t+1}(\Gamma_{t+1}+\Phi_t\Pi_t\Phi'_t)\Psi'_{t+1}\end{bmatrix}$$

and

$$K_t(z^t, u^t; z_{t+1} \in B)$$

(2.4.9)

$$\sim N[\Psi_{t+1}(\Lambda_t u_t + \Phi_t \hat{x}_t) \text{ , } R_{t+1} + \Psi_{t+1}(\Gamma_{t+1} + \Phi_t \Pi_t \Phi_t')\Psi_{t+1}'] \text{ .}$$

Proof: It has been noted that Theorem 2.3.2 applies. Thus the stochastic kernels G_t have densities g_t which can be computed from (2.3.17).

It will be assumed as an induction hypothesis that (2.4.4) holds for t where Π_t is non-singular. From (2.3.9), (2.4.1), and (2.4.2)

$$j_t(z^t, u^t; x_{t+1}, z_{t+1}) = \frac{1}{(2\pi)^{m/2}\sqrt{|R_{t+1}|}} \exp[-\tfrac{1}{2}(z_{t+1} - \Psi_{t+1}x_{t+1})'$$
$$\cdot R_{t+1}^{-1}(z_{t+1} - \Psi_{t+1}x_{t+1})] \int_{\mathcal{X}_t} \frac{1}{(2\pi)^{n/2}\sqrt{|\Gamma_{t+1}|}} \exp[-\tfrac{1}{2}(x_{t+1} - \Phi_t x_t - \Lambda_t u_t)'$$
$$\cdot \Gamma_{t+1}^{-1}(x_{t+1} - \Phi_t x_t - \Lambda_t u_t)] \frac{1}{(2\pi)^{n/2}\sqrt{|\Pi_t|}} \exp[-\tfrac{1}{2}(x_t - \hat{x}_t)\Pi_t^{-1}(x_t - \hat{x}_t)]dx_t \quad . \tag{2.4.10}$$

The integral over $\mathcal{X}_t$ will be computed first

$$\int_{\mathcal{X}_t} \exp -\tfrac{1}{2}[(x_{t+1} - \Phi_t x_t - \Lambda_t u_t)' \Gamma_{t+1}^{-1}(x_{t+1} - \Phi_t x_t - \Lambda_t u_t) + (x_t - \hat{x}_t)' \Pi_t^{-1}(x_t - \hat{x}_t)]dx_t$$

$$= \int_{\mathcal{X}_t} \exp -\tfrac{1}{2}[(x_{t+1} - \Lambda_t u_t)' \Gamma_{t+1}^{-1}(x_{t+1} - \Lambda_t u_t) + \hat{x}_t' \Pi_t^{-1}\hat{x}_t + x_t' \tilde{\Xi}^{-1} x_t - x_t'\tilde{\xi} - \tilde{\xi}' x_t]dx_t$$

(2.4.11)

$$= \int_{\mathcal{X}_t} \exp -\tfrac{1}{2}[(x_{t+1} - \Lambda_t u_t)' \Gamma_{t+1}^{-1}(x_{t+1} - \Lambda_t u_t) + \hat{x}_t'\Pi_t^{-1}\hat{x}_t - \tilde{\xi}'\tilde{\Xi}\tilde{\xi}$$
$$+ (x_t - \tilde{\Xi}\tilde{\xi})'\tilde{\Xi}^{-1}(x_t - \tilde{\Xi}\tilde{\xi})]dx_t$$

$$= (2\pi)^{n/2}\sqrt{\tilde{\Xi}} \; \exp -\tfrac{1}{2}[(x_{t+1} - \Lambda_t u_t)' \Gamma_{t+1}^{-1}(x_{t+1} - \Lambda_t u_t) + \hat{x}_t'\Pi_t^{-1}\hat{x}_t - \tilde{\xi}'\tilde{\Xi}\tilde{\xi}$$

where

$$\tilde{\Xi}^{-1} = \Phi_t'\Gamma_{t+1}^{-1}\Phi_t + \Pi_t^{-1} \tag{2.4.12}$$

and

$$\tilde{\xi} = \Phi_t'\Gamma_{t+1}^{-1}(x_{t+1} - \Lambda_t u_t) + \Pi_t^{-1}\hat{x}_t \text{ .} \tag{2.4.13}$$

$\tilde{\Xi}^{-1}$ is positive definite since $\Phi_t'\Gamma_{t+1}^{-1}\Phi_t$ is positive semi-definite and by assumption Π_t^{-1} is positive definite. The function $j_t(z^t, u^t; x_{t+1}, z_{t+1})$

will be examined as a function of x_{t+1} and z_{t+1} only. From (2.4.10), (2.4.11), and (2.4.13)

$$j_t(z^t,u^t;x_{t+1},z_{t+1}) = r_1 \exp-\tfrac{1}{2}[(z_{t+1}-\Psi_{t+1}x_{t+1})' R_{t+1}^{-1}(z_{t+1}-\Psi_{t+1}x_{t+1})$$

$$+ (x_{t+1}-\Lambda_t u_t)' \Gamma_{t+1}^{-1}(x_{t+1}-\Lambda_t u_t) - (\Phi_t'\Gamma_{t+1}^{-1}(x_{t+1}-\Lambda_t u_t) + \Pi_t^{-1}\hat{x}_t)' \tilde{\Sigma}$$

$$(2.4.14) \qquad \cdot (\Phi_t'\Gamma_{t+1}^{-1}(x_{t+1}-\Lambda_t u_t) + \Pi_t^{-1}\hat{x}_t)$$

$$= r_2 \exp-\tfrac{1}{2}\left[\begin{pmatrix} x_{t+1}-\Lambda_t u_t \\ z_{t+1}-\Psi_{t+1}\Lambda_t u_t \end{pmatrix}' \Sigma_{t+1}^{-1}\begin{pmatrix} x_{t+1}-\Lambda_t u_t \\ z_{t+1}-\Psi_{t+1}\Lambda_t u_t \end{pmatrix} - \begin{pmatrix} x_{t+1} \\ z_{t+1} \end{pmatrix}'\xi - \xi'\begin{pmatrix} x_{t+1} \\ z_{t+1} \end{pmatrix}\right]$$

$$= r_3 \exp-\tfrac{1}{2}\left[\begin{pmatrix} x_{t+1}-\Lambda_t u_t \\ z_{t+1}-\Psi_{t+1}\Lambda_t u_t \end{pmatrix} - \Sigma_{t+1}\xi\right]' \Sigma_{t+1}^{-1}\left[\begin{pmatrix} x_{t+1}-\Lambda_t u_t \\ z_{t+1}-\Psi_{t+1}\Lambda_t u_t \end{pmatrix} - \Sigma_{t+1}\xi\right],$$

where

$$(2.4.15) \qquad \Sigma_{t+1}^{-1} = \begin{bmatrix} \Psi_{t+1}'R_{t+1}^{-1}\Psi_{t+1} + \Gamma_{t+1}^{-1} - \Gamma_{t+1}^{-1}\Phi_t\tilde{\Sigma}\Phi_t'\Gamma_{t+1}^{-1} & -\Psi_{t+1}'R_{t+1}^{-1} \\ -R_{t+1}^{-1}\Psi_{t+1} & R_{t+1}^{-1} \end{bmatrix},$$

$$(2.4.16) \qquad \xi = \begin{pmatrix} \Gamma_{t+1}^{-1}\Phi_t\tilde{\Sigma}\Pi_t^{-1}\hat{x}_t \\ 0 \end{pmatrix},$$

and r_1, r_2, and r_3 are independent of x_{t+1} and z_{t+1}. From (2.4.14) j_t is a Gaussian density in (x_{t+1},z_{t+1}) with covariance Σ_{t+1} and mean

$$(2.4.17) \qquad \begin{pmatrix} \Lambda_t u_t \\ \Psi_{t+1}\Lambda_t u_t \end{pmatrix} + \Sigma_{t+1}\xi .$$

From (2.4.12) it can easily be shown by direct computation that

$$(2.4.18) \qquad \Gamma_{t+1}^{-1} - \Gamma_{t+1}^{-1}\Phi_t\tilde{\Sigma}\Phi_t'\Gamma_{t+1}^{-1} = (\Gamma_{t+1} + \Phi_t\Pi_t\Phi_t')^{-1} .$$

This identity will be used to compute Σ_{t+1} and $\Sigma_{t+1}\xi$ explicitly. First, from (2.4.15) and (2.4.18)

$$(2.4.19) \qquad \Sigma_{t+1}^{-1} = \begin{bmatrix} \Psi_{t+1}'R_{t+1}^{-1}\Psi_{t+1} + (\Gamma_{t+1} + \Phi_t\Pi_t\Phi_t')^{-1} & -\Psi_{t+1}'R_{t+1}^{-1} \\ -R_{t+1}^{-1}\Psi_{t+1} & R_{t+1}^{-1} \end{bmatrix}.$$

From (2.4.19) it can be verified directly that Σ_{t+1} satisfies (2.4.8). Since Π_t, $\tilde{\Sigma}$, and Γ_{t+1} are non-singular, from (2.4.12)

$$(2.4.20)\qquad \Phi_t = \Phi_t\Pi_t\tilde{\Sigma}^{-1}\tilde{\Sigma}\Pi_t^{-1} = \Phi_t\Pi_t(\Phi_t'\Gamma_{t+1}^{-1}\Phi_t + \Pi_t^{-1})\tilde{\Sigma}\Pi_t^{-1}$$

$$= (\Phi_t\Pi_t\Phi_t'\Gamma_{t+1}^{-1}\Phi_t + \Phi_t)\tilde{\Sigma}\Pi_t^{-1} = (\Phi_t\Pi_t\Phi_t' + \Gamma_{t+1})\Gamma_{t+1}^{-1}\Phi_t\tilde{\Sigma}\Pi_t^{-1} .$$

Thus from (2.4.16), (2.4.8) and (2.4.20)

$$(2.4.21)\qquad \Sigma_{t+1}\xi = \begin{pmatrix} (\Gamma_{t+1} + \Phi_t\Pi_t\Phi_t')\Gamma_{t+1}^{-1}\Phi_t\tilde{\Sigma}\Pi_t^{-1}\hat{x} \\ \Psi_{t+1}(\Gamma_{t+1} + \Phi_t\Pi_t\Phi_t')\Gamma_{t+1}^{-1}\Phi_t\tilde{\Sigma}\Pi_t^{-1}\hat{x} \end{pmatrix} = \begin{pmatrix} \Phi_t\hat{x} \\ \Psi_{t+1}\Phi_t\hat{x} \end{pmatrix} .$$

The mean of the Gaussian distribution j_t can now be computed from (2.4.17) and (2.4.21), and the result (2.4.7) is thus established. The result (2.4.9) follows from (2.4.7), (2.4.8), (2.3.12) and (2.3.13) of Theorem 2.3.2, and the properties of the normal distribution.

From Theorem 2.3.2, G_{t+1} has a density g_{t+1} that satisfies (2.3.17). For z^{t+1} and u^t fixed the denominator of (2.3.17) is fixed. Thus from (2.3.17) and (2.4.14), as a function of x_{t+1}, g_{t+1} has the form

$$(2.4.22)\qquad g_{t+1}(z^{t+1},u^t;x_{t+1})$$

$$= r_1^*\exp-\tfrac{1}{2}[(x_{t+1}-\Lambda_t u_t)'(\Sigma^*)^{-1}(x_{t+1}-\Lambda_t u_t) - x_{t+1}'\xi^* - (\xi^*)'x_{t+1}]$$

$$= r_2^*\exp-\tfrac{1}{2}[(x_{t+1}-\Lambda_t u_t - \Sigma^*\xi^*)'(\Sigma^*)^{-1}(x_{t+1}-\Lambda_t u_t - \Sigma^*\xi^*)]$$

where

$$(2.4.23)\qquad (\Sigma^*)^{-1} = \Psi_{t+1}'R_{t+1}^{-1}\Psi_{t+1} + \Gamma_{t+1}^{-1} - \Gamma_{t+1}^{-1}\Phi_t'\tilde{\Sigma}\Phi_t\Gamma_{t+1}^{-1} ,$$

$$(2.4.24)\qquad \xi^* = \Psi_{t+1}'R_{t+1}^{-1}(z_{t+1} - \Psi_{t+1}\Lambda_t u_t) + \Gamma_{t+1}^{-1}\Phi_t\tilde{\Sigma}\Pi_t^{-1}\hat{x}_t ,$$

and r_1^*, r_2^* do not depend on x_{t+1}. It follows from (2.4.22) that g_{t+1} is a Gaussian density in x_{t+1} with covariance Σ^* and mean

$$(2.4.25)\qquad \Lambda_t u_t + \Sigma^*\xi^* .$$

From (2.4.23) and the identity (2.4.18)

(2.4.26) $$(\Sigma^*)^{-1} = \Psi'_{t+1}R^{-1}_{t+1}\Psi_{t+1} + (\Gamma_{t+1} + \Phi_t\Pi_t\Phi'_t)^{-1} = \Pi^{-1}_{t+1}$$

so that (2.4.6) holds. By assumption Γ_{t+1} is positive definite. Thus $(\Gamma_{t+1} + \Phi_t\Pi_t\Phi'_t)$ and hence $(\Gamma_{t+1} + \Phi_t\Pi_t\Phi'_t)^{-1}$ is positive definite since $\Phi_t\Pi_t\Phi'_t$ is positive semi-definite. It follows then that $(\Sigma^*)^{-1}$ and hence Π_{t+1} is positive definite since $\Psi'_{t+1}R^{-1}_{t+1}\Psi_{t+1}$ is positive semi-definite. From (2.4.24), (2.4.26), (2.4.20) and (2.4.6)

(2.4.27)
$$\begin{aligned}
\Sigma^*\xi^* &= \Pi_{t+1}[\Psi_{t+1}R^{-1}_{t+1}(z_{t+1} - \Psi_{t+1}\Lambda_t u_t) + \Gamma^{-1}_{t+1}\Phi_t\tilde{\Sigma}\Pi_t^{-1}\hat{x}_t] \\
&= \Pi_{t+1}[\Psi_{t+1}R^{-1}_{t+1}(z_{t+1} - \Psi_{t+1}\Lambda_t u_t) + (\Phi_t\Pi_t\Phi'_t + \Gamma_{t+1})^{-1}\Phi_t\hat{x}_t] \\
&= \Pi_{t+1}[\Psi_{t+1}R^{-1}_{t+1}(z_{t+1} - \Psi_{t+1}\Lambda_t u_t) + (\Pi^{-1}_{t+1} - \Psi'_{t+1}R^{-1}_{t+1}\Psi_{t+1})\Phi_t\hat{x}_t] \\
&= \Phi_t\hat{x}_t + \Pi_{t+1}\Psi_{t+1}R^{-1}_{t+1}(z_{t+1} - \Psi_{t+1}\Lambda_t u_t - \Psi_{t+1}\Phi_t\hat{x}_t) = \hat{x}_{t+1} - \Lambda_t u_t
\end{aligned}$$

where $\hat{x}_{t+1}$ is defined by (2.4.5). The mean for g_{t+1} can now be computed from (2.4.25) and (2.4.27), and it follows that G_{t+1} satisfies (2.4.4) where $\hat{x}_{t+1}$ and Π_{t+1} satisfy (2.4.5) and (2.4.6). It has already been noted that Π_{t+1} is positive definite. The theorem follows by induction.

Chapter 3: Statistics Sufficient for Control

3.1 Introduction

This chapter is devoted to further development of distribution theory required for the control problem. Here data reduction will be accomplished by exploiting certain aspects of the loss function. The intention in defining a "sufficient" statistic is to reduce the amount of data to be used in computing an optimal control problem. Unless a sufficient statistic of few dimensions can be found, it is virtually impossible to solve the control problem in any practical sense.

In defining sufficient statistics, only measurability properties of the loss function will be considered. For given measurable functions $V_t(x_t,u^{t-1})$, the class of loss functions depending on V_t will be considered. The sufficiency of a statistic is defined with respect to this class of loss functions or simply with respect to the functions $V_t(x_t,u^{t-1})$. This idea is made precise in Definition 3.2.6. In Chapter 5 it is shown under certain additional assumptions that for loss functions of the form (1.2.32) with cost functions

$$c_t(x_t,u^{t-1}) = c_t^V(V_t(x_t,u^{t-1}),u_{t-1}) , \tag{3.1.1}$$

optimal control laws can be found which depend on the statistic Y_t provided it is sufficient for V_t.

The usual definition of a statistic as a measurable function on the sample space $Y_t(z^t)$ will be modified slightly. For the control problem, a statistic $Y_t(z^t,u^{t-1})$ at time t depends explicitly on the controls u^{t-1} used before time t as well as the data z^t available at time t. If a control law $\{u\}$ is specified, a statistic in the usual sense is obtained by

$$Y_t(z^t,u^{t-1}(z^{t-1})) , \tag{3.1.2}$$

but this will vary with the control law. The use of statistics in the control problem may accomplish another simplification of the problem. If, as is usually the case, the statistics Y_t have common range spaces $\mathcal{Y}_t = \mathcal{Y}$, then the domain space of the control functions $U_t(y)$ is this common space $\mathcal{Y}$ rather than the space Z^t which of necessity varies with t.

To be consistent with the point of view of Chapter 4, it would be more appropriate to define a statistic by

$$(3.1.3) \qquad Y_t(z^t, \{u^{t-1}\})$$

permitting dependence on the truncated law $\{u^{t-1}\}$ rather than only on the value of $\{u^{t-1}\}$ at z^t as in (3.1.2). The present definition is used because it appears to be adequate for the discrete time problem.

It should be pointed out that a sufficient statistic whether of the form (3.1.2) or (3.1.3) performs an important function, in addition to data reduction and simplification of the domain spaces of the control functions in that it permits a control function $U_t(Y_t)$ at time t to depend on the previous law $\{u^{t-1}\}$. This property is used in section 5.5 where certain laws in Y are shown to be universally optimal (Definitions 5.5.2 and 5.5.3). This point is discussed more fully in section 5.1.

The definition of a sufficient statistic used here differs only in details from that of Dynkin [7] and Hinderer [8]. The requirement (α) of Hinderer [8] (p. 36) corresponds to the Definition 3.2.6. This becomes clear in section 5.2 where the cost function $c_t^V(v_t, u_{t-1})$ is essentially replaced by the cost function $c_t^Y(y_t, u_{t-1})$ defined by (5.2.2) and made possible by the stochastic kernels postulated in Definition 3.2.6. The property (β) of Hinderer [8] (p. 36) ensures that there exist kernels K_t^Y which satisfy (3.2.2) of the definition of a sufficient statistic (Definition 3.2.2).

It is noted in section 3.2 that for control laws $\{U^Y\}$ (Definition 3.2.3), the process Y_t is a Markov process. Thus laws $\{U^Y\}$ correspond to Markov laws (plans) studied by Blackwell [2], Strauch [16], and Hinderer [8] for system variable Y_t. A Markovian dynamic programming model with complete information, state variable Y_t, and transition kernel K_t^Y

(of Definition 3.2.2) can always be defined. The optimality properties of the optimal Markov law obtained, for example, from Theorem 6.2 of Hinderer [8] for this model (and countable spaces) are slightly weaker than those obtained in Chapter 5 for $\{\hat{U}^Y\}$.

In section 3.2 definitions and elementary properties of sufficient statistics are given. Among the definitions is that of the identity statistic (Definition 3.2.7) which is shown to be sufficient for all loss functions of the form (1.2.32). This justifies the restriction of attention in Chapter 5 to systems with sufficient statistics.

In section 3.3 additional properties of the loss function are exploited in order to find explicit sufficient statistics. The example which motivates Theorem 3.3.1 of this section is the "limited fuel" problem. For example, let

$$L = x_T' Q_T x_T + \delta_a\left(\sum_{t=0}^{T-1} |u_t|\right) \tag{3.1.4}$$

where

$$\delta_a(u) = \begin{cases} 0 & \text{if } u \leq a \\ \infty & \text{if } u > a . \end{cases} \tag{3.1.5}$$

The constant a represents the fuel available, $\Sigma\,|u_t|$ the total fuel used, and the first term in (3.1.4) represents a squared miss distance. This problem is more difficult than that of Chapter 7 (7.4.1) which can be thought of as an approximation to this loss function (see the discussion in section 7.1). While no optimal law will be derived for this loss function, Theorem 3.3.1 provides a sufficient statistic for the problem.

Theorem 3.3.1 can be obtained by the alternative method of augmenting the system vector. Although this method will not be used here, it will be discussed briefly both to motivate the loss structure (3.3.1), (3.3.4) and because it is a method used extensively in the literature to show equivalence of models. Let $M_t(u^t)$ be a function on the control space which satisfies the recursive relation (3.3.1), and let the loss function have the form (1.2.32) with

$$c_t(x_t, u^{t-1}) = \tilde{c}_t(x_t, M_{t-1}(u^{t-1}), u_{t-1}) . \tag{3.1.6}$$

The loss function then clearly becomes Markov (c_t depends only on the most recent control vector u_{t-1}) if the system is replaced by the augmented system vector

$$\tilde{x}_t = (x_t, M_{t-1}) \ . \tag{3.1.7}$$

The system transition functions (1.2.2) for the augmented system

$$\tilde{x}_{t+1} = \tilde{\varphi}_t(\tilde{x}_t, u_t, w_{t+1}) \tag{3.1.8}$$

are obtained from (1.2.2) and (3.3.1)

$$\begin{aligned} \tilde{x}_{t+1} &= (x_{t+1}, M_t) \\ &= (\varphi_t(x_t, u_t, w_{t+1}), \bar{M}_t(M_{t-1}, u_t)) \\ &= \tilde{\varphi}_t(x_t, M_{t-1}, u_t, w_{t+1}) = \tilde{\varphi}_t(\tilde{x}_t, u_t, w_{t+1}) \ . \end{aligned} \tag{3.1.9}$$

By further augmenting the system process, the loss function can be reduced so that it depends only on the final system value. Let

$$\tilde{\tilde{x}}_t = (x_t, M_{t-1}, \ell_t) \tag{3.1.10}$$

where

$$\ell_t = \sum_{\tau=1}^{t} \tilde{c}_\tau(x_\tau, M_{\tau-1}(u^{\tau-1}), u_{\tau-1}) \ . \tag{3.1.11}$$

Then

$$L = \ell_T = L(\tilde{\tilde{x}}_T) \ , \tag{3.1.12}$$

and from (3.1.9) and (3.1.11)

$$\begin{aligned} \tilde{\tilde{x}}_{t+1} &= (x_{t+1}, M_t, \ell_{t+1}) \\ &= (x_{t+1}, M_t, \ell_t + \tilde{c}_{t+1}(x_{t+1}, M_t, u_t)) \\ &= (\tilde{\varphi}_t(x_t, M_{t-1}, u_t, w_{t+1}), \ell_t + \tilde{c}_{t+1}(\tilde{\varphi}_t(x_t, M_{t-1}, u_t, w_{t+1}), u_t)) \\ &= \tilde{\tilde{\varphi}}_t(x_t, M_{t-1}, \ell_t, u_t, w_{t+1}) = \tilde{\tilde{\varphi}}_t(\tilde{\tilde{x}}_t, u_t, w_{t+1}) \ . \end{aligned} \tag{3.1.13}$$

For the incomplete information problem considered here, reduction to a Markov loss function is accomplished by the introduction of sufficient statistics. Thus the method of augmented vectors is applied more appropriately to the structure and the sufficient statistic vectors rather than to the system vector. Results obtained by this method are given in section 3.3.

3.2 Definitions and preliminary results

It is assumed throughout this section that a control model is given. The class of laws $\mathcal{D}$ and the loss function L will be mentioned explicitly when they are required.

Definition 3.2.1. A measurable function $Y_t(z^t, u^{t-1})$ from $(Z^t \times \mathcal{U}^{t-1}, \mathcal{B}_Z^t \times \mathcal{B}_{\mathcal{U}}^{t-1})$ to $(\mathcal{Y}_t, \mathcal{B}_{\mathcal{Y}_t})$ for $t=1,2,\ldots$ will be called a statistic. For $t=0$, let

$$(3.2.1) \qquad Y_0 = y_0 \quad \text{where} \quad \{y_0\} = \mathcal{Y}_0 \; .$$

The definition (3.2.1) is consistent with the idea that Y_0 is a function of (z_0, u^{-1}) — that is, no data and no previous controls.

Definition 3.2.2. A statistic $Y_t(z^t, u^{t-1})$ is a sufficient statistic for the class $\mathcal{D}$ provided there exist stochastic kernels $K_t^Y(y_t, u_t; A)$ on $(\mathcal{Y}_t \times \mathcal{U}_t, \mathcal{Y}_{t+1})$ with the property

$$(3.2.2) \qquad P_{\{u\}}[Y_{t+1}(z^{t+1}, u^t(z^t)) \in A | z^t] = K_t^Y(Y_t(z^t, u^{t-1}(z^{t-1})), u_t(z^t); A) \quad \text{a.s. } P_{\{u\}}^t$$

for all control laws $\{u\} \in \mathcal{D}$ and $t=0,1,\ldots$.

The distributions K_t^Y are transitions for the statistic and are in a sense analogous to the transitions P_t with the important difference that Y_t is observable.

Definition 3.2.3. $\{U^Y\}$ is a control law in the statistic Y_t provided

$$(3.2.3) \qquad \{U^Y\} = \{U_0, U_1(\cdot), \ldots, U_{T-1}(\cdot)\}$$

where the control functions $U_t(y_t)$ are measurable functions from $\mathcal{Y}_t$ to $\mathcal{U}_t$.

A control law $\{u\}$ in the usual sense is obtained from a control law $\{U^Y\}$ in Y by successive substitution

$$(3.2.4) \qquad u_t(z^t) = U_t(Y_t(z^t, u^{t-1}(z^{t-1}))) \; , \quad t=0,1,\ldots,T-1 \; .$$

For a control law obtained in this manner the sufficient statistic Y_t is a Markov process with transitions K_t^Y. For laws that do not satisfy (3.2.4), Y_t will in general not be Markov. It will be seen later that there is strong motivation for restricting attention to controls of the type (3.2.4).

The transitions K_t^Y are related to the kernels $K_t(z^t,u^t;B)$ defined by (2.2.13). Conditional on z^t the distribution of $Y_{t+1}(z^{t+1},u^t)$ is induced by that of z_{t+1}.

Lemmas 3.2.1, 3.2.2, and 3.2.3 require the existence of stochastic kernels G_t and K_t which satisfy (2.1.1) and (2.1.2) for all $\{u\} \in \mathfrak{D}$. Existence of such kernels is assured by Theorem 2.2.3 under the assumption that the spaces $\mathcal{X}_t$ and Z_t are finite dimensional Euclidean spaces, and by Theorem 2.3.1 under the assumption of the existence of densities for the kernels H_t.

Lemma 3.2.1. Let $Y_t(z^t,u^{t-1})$ be a statistic and suppose there exists stochastic kernels $K_t^Y(y_t,u_t;A)$ which satisfy

$$K_t^Y(Y_t(z^t,u^{t-1}),u_t;A) = K_t(z^t,u^t;\{z_{t+1} \mid Y_{t+1}(z^t,z_{t+1},u^t) \in A\}) \tag{3.2.5}$$

for all (z^t,u^t), $A \in \mathcal{B}_{\mathcal{Y}_{t+1}}$ and $t=1,2,\ldots$ where $K_t(z^t,u^t;B)$ satisfies (2.1.2) for all $\{u\} \in \mathfrak{D}$. Then the statistic Y_t is sufficient for $\mathfrak{D}$.

Proof: From (3.2.5), (2.1.2), and Lemma A.1.5 for $\{u\} \in \mathfrak{D}$

$$K_t^Y(Y_t(z^t,u^{t-1}(z^{t-1})),u_t(z^t);A)$$

$$= \int \chi_A(Y_{t+1}(z^t,z_{t+1},u^t(z^t)))K_t(z^t,u^t(z^t),dz_{t+1})$$

$$= E_{\{u\}}[\chi_A(Y_{t+1}(z^{t+1},u^t(z^t)))\mid z^t]$$

$$= P_{\{u\}}[Y_{t+1}(z^{t+1},u^t(z^t)) \in A \mid z^t] \qquad \text{a.s. } P_{\{u\}}^t .$$

Thus (3.2.2) holds and Y_t is a sufficient statistic.

Next, the relationship of sufficient statistics to the loss function will be explored.

Sufficiency of a statistic for a particular loss function will depend only on measurability properties of the cost functions $c_t(x_t, u^{t-1})$. To make this precise the idea of loss structure V_t is introduced.

Definition 3.2.4. A loss structure V consists of measurable functions $V_t(x_t, u^{t-1})$

(3.2.6) $$V_t : (\mathcal{X}_t \times \mathcal{U}^{t-1}, \mathcal{B}_{\mathcal{X}_t} \times \mathcal{B}_{\mathcal{U}}^{t-1}) \longrightarrow (\mathcal{V}_t, \mathcal{B}_{\mathcal{V}_t})$$

$t = 1, 2, \ldots$.

Definition 3.2.5. A loss function L has structure V provided there exist non-negative measurable cost functions $c_t^V(v_t, u_{t-1})$ such that

(3.2.7) $$L(x^T, u^{T-1}) = \sum_{t=1}^{T} c_t^V(V_t(x_t, u^{t-1}), u_{t-1})$$

for all $(x^T, u^{T-1}) \in \mathcal{X}^T \times \mathcal{U}^{T-1}$.

A statistic Y_t is said to be sufficient for a structure V_t provided the conditional distribution of V_t given z^t depends on the data z^t and the previous controls u^{t-1} only thru $Y_t(z^t, u^{t-1})$.

Definition 3.2.6. A statistic Y is sufficient for a loss structure V provided Y is a sufficient statistic and in addition there exist stochastic kernels $G_t^{Y,V}(y_t, A)$ on $(\mathcal{Y}_t, \mathcal{V}_t)$ that satisfy

(3.2.8) $$P_{\{u\}}[V_t(x_t, u^{t-1}(z^{t-1})) \in A | z^t] = G_t^{Y,V}(Y_t(z^t, u^{t-1}(z^{t-1})); A) \quad \text{a.s. } P_{\{u\}}$$

for all $\{u\} \in \mathcal{D}$, $A \in \mathcal{B}_{\mathcal{V}_t}$, $t = 1, 2, \ldots$. A statistic Y is sufficient for the control system provided L has structure V and Y is sufficient for V .

The kernels $G_t^{Y,V}$ are related to the kernels G_t which satisfy (2.1.2) since the distribution of $V_t(x^t, u^{t-1})$ conditional on z^t is induced by that of x_t .

Lemma 3.2.2. Let Y_t be a sufficient statistic and suppose there exist stochastic kernels $G_t^{Y,V}(y_t, A)$ which satisfy

(3.2.9) $$G_t^{Y,V}(Y_t(z^t, u^{t-1}), A) = G_t(z^t, u^{t-1}; \{x_t | V_t(x_t, u^{t-1}) \in A\})$$

for all $A \in \mathcal{B}_{\mathcal{V}_t}$, (z^t, u^{t-1}) , and $t = 1, 2, \ldots$ where V is a loss structure and G_t is a stochastic kernel which satisfies (2.1.1) for all $\{u\} \in \mathcal{D}$. Then Y is sufficient for V .

Proof: From (3.2.9), (2.1.1), and Lemma A.1.5 for $\{u\} \in \mathcal{D}$

$$G_t^{Y,V}(Y_t(z^t,u^{t-1}(z^{t-1})),A) = \int \chi_A(V_t(x_t,u^{t-1}(z^{t-1})))G_t(z^t,u^t(z^t);dx_t)$$

$$= E_{\{u\}}[\chi_A(V_t(x_t,u^{t-1}(z^{t-1}))|z^t]$$

$$= P_{\{u\}}[V_t(x_t,u^{t-1}(z^{t-1})) \in A|z^t] \qquad \text{a.s. } P_{\{u\}} .$$

Thus (3.2.8) holds, and Y is sufficient for V .

Next, an essentially vacuous statistic and structure will be introduced. They are in a sense the most general possible and are of theoretical rather than practical interest.

Definition 3.2.7. The identity statistic I is defined by

$$I_t(z^t,u^{t-1}) = (z^t,u^{t-1}) . \tag{3.2.10}$$

It's range and domain space are the same, $(Z^t \times U^{t-1})$.

Definition 3.2.8. The identity loss structure I is defined by

$$I_t(x_t,u^{t-1}) = (x_t,u^{t-1}) . \tag{3.2.11}$$

The range and domain are the same, $(\mathcal{X}_t,U^{t-1})$.

For the identity structure a trivial result will be stated.

Lemma 3.2.3. All loss functions of the form (1.2.32) have identity structure.

Lemma 3.2.4. If there exist stochastic kernels G_t and K_t which satisfy (2.1.1) and (2.1.2) for $t=1,2,\ldots$, then the identity statistic is sufficient for the identity structure and

$$K_t^I(z^t,u^{t-1},u_t;A) = K_t(z^t,u^t;A_{(z^t,u^t)}) \tag{3.2.12}$$

$$G_t^{I,I}(z^t,u^{t-1};B) = G_t(z^t,u^{t-1};B_{(u^{t-1})}) \tag{3.2.13}$$

where $A \in \mathcal{B}_Z^{t+1} \times \mathcal{B}_u^t = \mathcal{B}_{Z_{t+1}} \times \mathcal{B}_Z^t \times \mathcal{B}_u^t$, $B \in \mathcal{B}_{X_t} \times \mathcal{B}_u^{t-1}$, $A_{(z^t,u^t)}$ is the section of A at (z^t,u^t) and $B_{(u^{t-1})}$ is the section of B at u^{t-1} .

Proof: Since K_t , G_t , $\chi_A(z^{t+1},u^t)$, and $\chi_B(x_t,u^{t-1})$ are stochastic kernels, from (3.2.12), (3.2.13) and Lemma A.1.2 ii)

$$K_t^I(z^t,u^{t-1},u_t;A) = \int \chi_A(z^{t+1},u^t)K_t(z^t,u^t;dz_{t+1})$$

$$G_t^{I,I}(z^t,u^{t-1};B) = \int \chi_B(x_t,u^{t-1})\,G_t(z^t,u^{t-1};dx_t)$$

are stochastic kernels. The results then follow from Lemmas 3.2.1 and 3.2.2.

For each control law $\{u\}$, a statistic Y_t induces a distribution on $(\mathcal{Y}_t,\mathcal{B}_{\mathcal{Y}_t})$.

Definition 3.2.9. Let $\{u\}$ be a control law and $Y_t(z^t,u^{t-1})$ a statistic, then $(Y^{-1}P)^t_{\{u\}}$ is a probability measure on $(\mathcal{Y}_t,\mathcal{B}_{\mathcal{Y}_t})$ defined by

$$(Y^{-1}P)^t_{\{u\}}(B) = P^t_{\{u\}}[\{z^t|Y^t(z^t,u^{t-1}(z^{t-1})) \in B\}] \tag{3.2.14}$$

for $B \in \mathcal{B}_{\mathcal{Y}_t}$ where $P^t_{\{u\}}$ is given by (1.2.20).

It should be noticed that $(Y^{-1}P)^t_{\{u\}}$ depends only on the truncated control law $\{u^{t-1}\}$.

Definition 3.2.10. Let Y_t be a statistic, let

$$\rho = \{\rho_1,\rho_2,\dots\} \tag{3.2.15}$$

be σ-finite measures on the spaces $(\mathcal{Y}_t,\mathcal{B}_{\mathcal{Y}_t})$, $t=1,2,\dots$, and let $\mathcal{D}$ be a class of admissible control laws. Then define

$$\mathcal{D}_\rho = \{\{u\}|\{u\} \in \mathcal{D},\ (Y^{-1}P)^t_{\{u\}} << \rho_t,\ t=1,2,\dots\} \tag{3.2.16}$$

[$(Y^{-1}P)^t_{\{u\}} << \rho_t$ indicates that the measure $(Y^{-1}P)^t_{\{u\}}$ is absolutely continuous with respect to the measure ρ_t . (see e.g. Doob [4] p. 611)].

Definition 3.2.11. A statistic Y is dominated by ρ provided

$$\mathcal{D}_\rho = \mathcal{D} . \tag{3.2.17}$$

Lemma 3.2.5. Let Y_t be a sufficient statistic for which

(3.2.18) $$K_t^Y(y_t,u_t;\cdot) \ll \rho_{t+1}$$

for all (y_t,u_t) and $t=0,1,\ldots$, then ρ dominates Y.

Proof: Take $N \in \mathcal{B}_{\mathcal{Y}_{t+1}}$ such that $\rho_{t+1}(N)=0$, then from (3.2.18) $K_t^Y(y_t,u_t;N)=0$ for all y_t,u_t. Thus from (3.2.2)

$$P_{\{u\}}[Y_{t+1}(z^{t+1},u^t(z^t)) \in N|z^t] = 0 \qquad \text{a.s. } P^t_{\{u\}} .$$

Thus from (3.2.14)

$$(Y^{-1}P)^{t+1}_{\{u\}}(N) = P_{\{u\}}[Y_{t+1}(z^{t+1},u^t(z^t) \in N] = 0 .$$

It follows then that

$$(Y^{-1}P)^t_{\{u\}} \ll \rho_{t+1}$$

and hence from (3.2.16) that (3.2.17) holds.

It may be noted that in general the identity statistic will not be dominated.

Lemma 3.2.6. If the spaces $\mathcal{U}_t$ are non-countable and the constant controls

(3.2.19) $$u_t(z^t) = u^* \qquad \forall\ z^t$$

are admissible, then the identity statistic I is not dominated.

Proof: Let $\{u\}$ be a control law for which (3.2.19) holds, where, of course, $u^* \in \mathcal{U}_t$, then

$$(I^{-1}P)^{t+1}_{\{u\}}(Z^{t+1}\times\mathcal{U}^{t-1}\times\{u^*\})=P_{\{u\}}[I_{t+1}(z^{t+1},u^{t-1},u^t)\in Z^{t+1}\times\mathcal{U}^{t-1}\times\{u^*\}] =1 .$$

Thus if ρ_t dominates I, it follows from Definition 3.2.11 that

$$\rho_{t+1}(Z^{t+1}\times\mathcal{U}^{t-1}\times\{u^*\}) \neq 0$$

for all singletone $\{u^*\}$ in $\mathcal{U}_t$. Since $\mathcal{U}_t$ is non-countable, it follows that ρ_{t+1} cannot be σ-finite.

In order to avoid cumbersome notation involving sufficient statistics and structures the following notational conventions will be observed. When the control law $\{u\}$ involved is clear

$$Y_t = Y_t(z^t, u^{t-1}(z^{t-1}))$$

$$(3.2.20) \quad \begin{aligned} V_t &= V_t(x_t, u^{t-1}(z^{t-1})) \\ u_t &= u_t(z^t) \\ c_t^V &= c_t^V(V_t(x_t, u^{t-1}(z^{t-1})), u_{t-1}(z^{t-1})) \; . \end{aligned}$$

If the control law is given some designation such as $\{\bar{u}\}$ or $\{\hat{u}\}$ then the same designation will be given to Y_t, V_t, u_t, and c_t^V. For example,

$$(3.2.21) \qquad \hat{Y}_t = Y_t(z^t, \hat{u}^{t-1}(z^{t-1})) \; .$$

3.3 Equivalent statistics

This section is devoted to results which are useful in the identification of sufficient statistics.

First, additional structure will be defined on the control space $\mathcal{U}^t$. Let $M_t(u^t)$ be a measurable function from $(\mathcal{U}^t, \mathcal{B}_{\mathcal{U}}^t)$ to the measurable space $(\mathcal{M}_t, \mathcal{B}_{\mathcal{M}_t})$ and let $\bar{M}_t(m_{t-1}, u_t)$ be defined from $\mathcal{M}_{t-1} \times \mathcal{U}_t$ to $\mathcal{M}_t$ and satisfy

$$(3.3.1) \qquad M_t(u^t) = \bar{M}_t(M_{t-1}(u^{t-1}), u_t)$$

for all $u^t \in \mathcal{U}^t$. For example, let

$$(3.3.2) \qquad M_t(u^t) = \sum_{\tau=0}^{t} \bar{c}_\tau(u_\tau) \; .$$

Then

$$(3.3.3) \qquad \bar{M}_t(m_{t-1}, u_t) = m_{t-1} + \bar{c}_t(u_t) \; .$$

Theorem 3.3.1. Let Y_t be a statistic sufficient for V_t and let

$$(3.3.4) \qquad \tilde{V}_t(x_t, u^{t-1}) = \bar{V}_t(V_t(x_t, u^{t-1}), M_{t-1}(u^{t-1}))$$

where $\bar{V}_t(v_t, m_{t-1})$ is a measurable function from $\mathcal{V}_t \times \mathcal{M}_{t-1}$ to $\tilde{\mathcal{V}}_t$ and

$M_t(u^t)$ satisfies (3.3.1). Then the statistic

$$(3.3.5) \qquad \tilde{Y}_t(z^t,u^{t-1}) = (Y_t(z^t,u^{t-1}),M_{t-1}(u^{t-1}))$$

is sufficient for $\tilde{V}_t$.

Proof: Since Y_t is sufficient for V_t from Definitions 3.2.2 and 3.2.6 there exists stochastic kernels K_t^Y and $G_t^{Y,V}$ which satisfy (3.2.2) and (3.2.8). Define $\tilde{K}_t$ and $\tilde{G}_t$ by

$$(3.3.6) \qquad \tilde{K}_t(y_t,m_{t-1},u_t;A) = K_t^Y(y_t,u_t;\{y_{t+1}|(y_{t+1},\bar{M}_t(m_{t-1},u_t) \in A\})$$

for $A \in \mathcal{B}_{\mathcal{Y}_{t+1}} \times \mathcal{B}_{\mathcal{M}_t}$ and

$$(3.3.7) \qquad \tilde{G}_t(y_t,m_{t-1};B) = G_t^{Y,V}(y_t;\{v_t|\bar{V}_t(v_t,m_{t-1}) \in B\})$$

for $B \in \mathcal{B}_{\tilde{\mathcal{V}}_t}$. From Lemma A.1.4 $\tilde{K}_t$ is a stochastic kernel on

$$(\mathcal{Y}_t \times \mathcal{M}_{t-1} \times \mathcal{U}_t, \mathcal{Y}_{t+1} \times \mathcal{M}_t) = (\tilde{\mathcal{Y}}_t \times \mathcal{U}_t, \tilde{\mathcal{Y}}_{t+1})$$

and $\tilde{G}_t$ is a stochastic kernel on

$$(\mathcal{Y}_t \times \mathcal{M}_{t-1}, \tilde{\mathcal{V}}_t) = (\tilde{\mathcal{Y}}_t, \tilde{\mathcal{V}}_t) .$$

For a given control law $\{u\}$ from (3.3.5), (3.3.6), (3.3.1), (3.2.2), Lemma A.1.5, and (3.3.5) again

$$\tilde{K}_t(\tilde{Y}_t(z^t,u^{t-1}(z^{t-1})),u_t(z^t);A) = \tilde{K}_t(Y_t(z^t,u^{t-1}(z^{t-1})),M_{t-1}(u^{t-1}(z^{t-1})),u_t(z^t);A)$$

$$= \int \chi_A(y_{t+1},\bar{M}_t(M_{t-1}(u^{t-1}(z^{t-1})),u_t(z^t))K_t(Y_t(z^t,u^{t-1}(z^{t-1})),u_t(z^t);dy_{t+1})$$

$$= \int \chi_A(y_{t+1},M_t(u^t(z^t)))K_t(Y_t(z^t,u^{t-1}(z^{t-1})),u_t(z^t);dy_{t+1})$$

$$= E_{\{u\}}[\chi_A(Y_{t+1}(z^{t+1},u^t(z^t)),M_t(u^t(z^t)))|z^t]$$

$$= E_{\{u\}}[\tilde{Y}_{t+1} \in A|z^t] \qquad \text{a.s. } P^t_{\{u\}} .$$

Thus (3.2.2) holds for $K_t^{\tilde{Y}} = \tilde{K}_t$, and it follows that $\tilde{Y}_t$ is a sufficient statistic. Similarly, from (3.3.5), (3.3.7), (3.2.8), Lemma A.1.5, and (3.3.4)

$$\tilde{G}_t(\tilde{Y}_t(z^t,u^{t-1}(z^{t-1}));B) = \tilde{G}_t(Y_t(z^t,u^{t-1}(z^{t-1})),M_{t-1}(u^{t-1}(z^{t-1}));B)$$

$$= \int \chi_B(\bar{V}_t(v_t,M_{t-1}(u^{t-1}(z^{t-1}))))G_t^{Y,V}(Y_t(z^t,u^{t-1}(z^{t-1})),dv_t)$$

$$= E_{\{u\}}[\chi_B(\bar{V}_t(V_t(x_t,u^{t-1}(z^{t-1})),M_{t-1}(u^{t-1}(z^{t-1}))))|z^t]$$

$$= P_{\{u\}}[\tilde{V}_t \in B|z^t] \qquad \text{a.s. } P_{\{u\}}^t .$$

Thus (3.2.8) holds for $G_t^{\tilde{Y},\tilde{V}} = \tilde{G}_t$, and $\tilde{Y}_t$ is sufficient for $\tilde{V}_t$.

Corollary 3.3.1. Let Y_t be a statistic sufficient for V_t , and let

(3.3.8) $$\tilde{V}_t(x_t,u^{t-1}) = \bar{V}_t(V_t(x_t,u^{t-1}), \sum_{\tau=0}^{t-1} \bar{c}_\tau(u_\tau)) .$$

Then

(3.3.9) $$\tilde{Y}_t = (Y_t, \sum_{\tau=0}^{t-1} \bar{c}_\tau(u_\tau))$$

is sufficient for $\tilde{V}_t$.

Proof: This result follows from Theorem 3.3.1 for M_t and $\bar{M}_t$ given by (3.3.2) and (3.3.3).

A statistic Y_t is equivalent to the filtering distributions $G_t(z^t,u^{t-1};A)$ provided it acts as a parameter for the family of distributions G_t indexed by (z^t,u^{t-1}) and further it can be recovered from G_t . Thus knowledge of Y_t is equivalent to knowledge of the filtering distribution G_t .

Definition 3.3.1. Let the spaces $\mathcal{X}_t$, $t=0,1,2,\ldots$ and $\mathcal{Z}_t$, $t=1,2,\ldots$ be finite dimensional Euclidean spaces, and let G_t be the filtering distribution assured by Theorem 2.2.3. Then the statistic Y_t is equivalent to the filtering distribution provided there exists stochastic kernels $G_t^Y(y_t,A)$ on $(\mathcal{Y}_t,\mathcal{X}_t)$ which satisfy

(3.3.10) $$G_t(z^t,u^{t-1};A) = G_t^Y(Y_t(z^t,u^{t-1});A)$$

for all (z^t,u^{t-1}) , $A \in \mathcal{B}_{\mathcal{X}_t}$ and $t=1,2,\ldots$ and further there exists

(3.3.11) $$\theta_t : (\mathcal{X}_t^*, \mathcal{B}_{\mathcal{X}_t^*}) \longrightarrow (\mathcal{Y}_t, \mathcal{B}_{\mathcal{Y}_t})$$

measurable such that

(3.3.12) $$\theta_t(G_t^Y(y;\cdot)) = y$$

for all $y \in \mathcal{Y}_t$. $\mathcal{X}_t^*$ indicates the space of all probability measures on $(\mathcal{X}_t, \mathcal{B}_{\mathcal{X}_t})$ and $\mathcal{B}_{\mathcal{X}_t^*}$ the σ-field generated by $G(A)$ for $A \in \mathcal{B}_{\mathcal{X}_t}$.

<u>Theorem</u> 3.3.2. Let the spaces $\mathcal{X}_t$ and Z_t be finite dimensional Euclidean spaces, and let Y_t be equivalent to the filtering distributions. Then there exists $\overline{Y}_{t+1}(y_t, z_{t+1}, u_t)$ measurable which satisfies

$$\text{(3.3.13)} \qquad \overline{Y}_{t+1}(Y_t(z^t, u^{t-1}), z_{t+1}, u_t) = Y_{t+1}(z^{t+1}, u^t) \qquad \text{a.s. } K_t(z^t, u^t; B)$$

for all (z^t, u^t) where K_t is determined as in Theorem 2.2.3. Further Y_t is sufficient for $V_t(x_t, u^{t-1}) = x_t$.

Proof: From Theorem 2.2.3, let K_t and G_t be stochastic kernels which satisfy (2.1.1), (2.1.2), (2.2.12), (2.2.13), and (2.2.17) - (2.2.19). Let $G_t^Y(y, A)$ satisfy (3.3.10). Then define

$$\text{(3.3.14)} \qquad J_{t,A}^*(y_t, u_t; B) = \int_{\mathcal{X}_t} P_t(x_t, u_t; A \times B)\, G_t^Y(y_t; dx_t)$$

$$\text{(3.3.15)} \qquad K_t^*(y_t, u_t; B) = J_{t,\mathcal{X}_{t+1}}^*(y_t, u_t; B) .$$

From Lemma 2.2.3, there exists a stochastic kernel $G_{t+1}^*(y_t, z_{t+1}, u_t; A)$ which satisfies

$$\text{(3.3.16)} \qquad G_{t+1}^*(y_t, z_{t+1}, u_t; A) = \frac{J_{t,A}^*(y_t, u_t; dz_{t+1})}{K_t^*(y_t, u_t; dz_{t+1})} \qquad \text{a.s. } K_t^*(y_t, u_t; B) .$$

Define $\overline{Y}_{t+1}$ by

$$\text{(3.3.17)} \qquad \overline{Y}_{t+1}(y_t, z_{t+1}, u_t) = \theta_{t+1}(G_{t+1}^*(y_t, z_{t+1}, u_t; \cdot)) .$$

Since G_{t+1}^* is a stochastic kernel, from Lemma A.1.12 $\mathcal{G}^*$ defined by $\mathcal{G}^*(y_t, z_{t+1}, u_t) = G_{t+1}^*(y_t, z_{t+1}, u_t; \cdot)$ is a measurable transformation from $\mathcal{Y}_t \times Z_{t+1} \times \mathcal{U}_t$ to $\mathcal{X}_t^*$. Thus $\overline{Y}_{t+1}$ defined by (3.3.17) is a measurable transformation from $\mathcal{Y}_t \times Z_{t+1} \times \mathcal{U}_t$ to $\mathcal{Y}_{t+1}$.

From (2.2.12), (2.2.18), (2.2.19), (3.3.10), (3.3.14), and (3.3.15)

$$\text{(3.3.18)} \qquad J_{t,A}(z^t, u^t; B) = J_{t,A}^*(Y_t(z^t, u^{t-1}), u_t; B)$$

$$K_t(z^t, u^t; B) = K_t^*(Y_t(z^t, u^{t-1}), u_t; B)$$

for all (z^t, u^t) , $B \in \mathcal{B}_{Z_{t+1}}$. Further from (3.3.16)

$$\int_B G^*_{t+1}(Y_t(z^t,u^t), z_{t+1}, u_t; A) K^*_t(Y_t(z^t,u^{t-1}), u_t; dz_{t+1})$$
$$= J^*_{t,A}(Y_t(z^t,u^{t-1}), u_t; B)$$

for all B . Thus from (3.3.18)

$$\int_B G^*_{t+1}(Y_t(z^t,u^t), z_{t+1}, u_t; A) K_t(z^t,u^t; dz_{t+1})$$
$$= J_{t,A}(z^t,u^t; B)$$

for all B . Thus

$$G^*_{t+1}(Y_t(z^t,u^{t-1}), z_{t+1}, u_t; A) = \frac{J_{t,A}(z^t,u^t; dz_{t+1})}{K_t(z^t,u^t; dz_{t+1})}(z_{t+1}) \text{ a.s. } K_t(z^t,u^t; B) . \tag{3.3.19}$$

Thus from (3.3.10) for $t+1$, (2.2.17), and (3.3.19)

$$G^*_{t+1}(Y_t(z^t,u^{t-1}), z_{t+1}, u_t; A) = G^Y_{t+1}(Y_{t+1}(z^{t+1},u^t); A) \quad \text{a.s. } K_t(z^t,u^t; B) \tag{3.3.20}$$

for all (z^t,u^t) and $A \in \mathcal{B}_{\mathcal{X}_{t+1}}$. For (z^t,u^t) fixed the exceptional set in z_{t+1} depends on A . Since both sides of (3.3.20) are stochastic kernels on $(Z_{t+1}, \mathcal{X}_{t+1})$ for (z^t,u^t) fixed, from Lemma A.1.11

$$K_t(z^t,u^t; \{z_{t+1} \mid G^*_{t+1}(Y_t(z^t,u^{t-1}), z_{t+1}, u_t; A)$$
$$= G^Y_{t+1}(Y_{t+1}(z^{t+1},u^t); A) \quad \text{for all } A \in \mathcal{B}_{\mathcal{X}_{t+1}}\}) = 1$$

for all z^t, u^t . For $z_{t+1} \in \{\text{above}\}$ from (3.3.17) and (3.3.12)

$$\overline{Y}_{t+1}(Y_t(z^t,u^{t-1}), z_{t+1}, u_t) = \theta_{t+1}(G^*_{t+1}(Y_t(z^t,u^{t-1}), z_{t+1}, u_t; \cdot)) =$$
$$= \theta_{t+1}(G^Y_{t+1}(Y_{t+1}(z^{t+1},u^t); \cdot)) = Y_{t+1}(z^{t+1},u^t) .$$

Thus (3.3.13) holds. Let

$$K^Y_t(y_t,u_t; C) = K^*_t(y_t,u_t; \{z_{t+1} \mid \overline{Y}_{t+1}(y_t, z_{t+1}, u_t) \in C\}) . \tag{3.3.21}$$

Since K^*_t is a stochastic kernel and $\overline{Y}_{t+1}$ is measurable, from Lemma A.1.4, K^Y_t is a stochastic kernel. From (3.3.21), (3.3.18), and (3.3.13)

$$K_t^Y(Y_t(z^t,u^{t-1});C) = K_t^*(Y_t(z^t,u^{t-1}),u_t;\ \{z_{t+1}|\bar{Y}_{t+1}(Y_t(z^t,u^{t-1}),z_{t+1},u_t)\in C\})$$

$$= K_t(z^t,u^t;\ \{z_{t+1}|\bar{Y}_{t+1}(Y_t(z^t,u^{t-1}),z_{t+1},u_t)\in C\})$$

$$= K_t(z^t,u^t;\ \{z_{t+1}|Y_{t+1}(z^{t+1},u^t)\in C\})\ .$$

Thus (3.2.5) holds and from Lemma 3.2.1 Y_t is a sufficient statistic. Let

$$G_t^{Y,V}(y_t,A) = G_t^Y(y_t,A)\ .$$

Then from (3.3.10)

$$G_t^{Y,V}(Y_t(z^t,u^{t-1}),A) = G_t(z^t,u^t;A)$$
$$= G_t(z^t,u^{t-1};\ \{x_t|V_t(x_t,u^{t-1})\in A\})\ .$$

Thus (3.2.9) holds and from Lemma 3.2.2, Y_t is sufficient for $V_t = x_t$.

Corollary 3.3.2. Let the spaces $\mathfrak{X}_t, Z_t$ be finite dimensional Euclidean spaces, let Y_t be a statistic equivalent to the filtering distributions, and let

(3.3.22) $$\tilde{V}(x_t,u^{t-1}) = \bar{V}_t(x_t,M_{t-1}(u^{t-1}))$$

where $M_t(u^t)$ satisfies (3.3.1). Then

(3.3.23) $$\tilde{Y}_t = (Y_t,m_{t-1})$$

is sufficient for $\tilde{V}$.

Proof: This follows from Theorems 3.3.1 and 3.3.2.

3.4 Sufficient statistics for linear Gaussian systems

From Theorem 2.4.1 for the linear Gaussian model, the filtering distribution $G_t(z^t,u^{t-1};A)$, is Gaussian with mean $\hat{x}_t(z^t,u^{t-1})$ and covariance Π_t which satisfy the recursive relations (2.4.5) and (2.4.6). This result will be used to obtain sufficient statistics for some simple loss structures.

Theorem 3.4.1. For the linear Gaussian model (1.3.1)-(1.3.5)

(3.4.1) $$Y_t(z^t,u^{t-1}) = \hat{x}_t(z^t,u^{t-1})$$

where $\hat{x}_t$ is computed by (2.4.5) and (2.4.6) is equivalent to the filtering distribution.

Proof: Let

$$(3.4.2) \qquad G_t^{\hat{x}}(\hat{x}_t;A) = \int_A \frac{1}{(2\pi)^{n/2}\sqrt{|\Pi_t|}} \exp -\tfrac{1}{2}(x_t-\hat{x}_t)'\Pi_t^{-1}(x_t-\hat{x}_t)dx_t$$

Then (3.3.10) of Definition 3.3.1 follows from (2.4.4) of Theorem 2.4.1. The function $\theta_t(G_t(\cdot))$ is defined by

$$(3.4.3) \qquad \theta_t(G_t(\cdot)) = \begin{cases} \int_{\mathcal{X}_t} x_t G_t(dx_t) & \text{if } \int_{\mathcal{X}_t} |x| G_t(dx_t) < \infty \\ 0 & \text{if } \int_{\mathcal{X}_t} |x_t| G_t(dx_t) = \infty \end{cases} .$$

From the properties of the Gaussian distribution $G_t^{\hat{x}}$ defined by (3.4.2) satisfies

$$(3.4.4) \qquad \hat{x}_t = \int_{\mathcal{X}_t} x_t G_t(\hat{x}_t;dx_t) .$$

Thus (3.3.12) of Definition 3.3.1 also holds. Approximating integrals as limits of integrals of simple functions, the measurability of θ_t given by (3.4.3) is easily established.

Since $\hat{x}_t$ is equivalent, from Theorem 3.3.2 there exists a function $\bar{x}_t(\hat{x}_t, z_{t+1}, u_t)$ which satisfies

$$(3.4.5) \qquad \hat{x}_{t+1} = \bar{x}_t(\hat{x}_t, z_{t+1}, u_t) \qquad \text{a.s. } K_t(z^t, u^t; B) .$$

This function is given explicitly by (2.4.5) of Theorem 2.4.1.

From Theorems 3.4.1, 3.3.2, and 3.3.1 sufficient statistics can be obtained for a wide variety of loss structures. Some of these will be given in Theorem 3.4.4. First, a sufficient statistic will be found for a structure, that is considered in the examples of Chapter 6 and 7.

It follows from Theorem 3.3.2 that $\hat{x}_t$ is sufficient for x_t. However, since the kernels $K_t^{\hat{x}}$ and $G_t^{\hat{x},x}$ will be required later, this result will be proved directly from Lemmas 3.2.1 and 3.2.2.

<u>Theorem</u> 3.4.2. For the linear Gaussian system, the statistic $\hat{x}_t(z^t, u^{t-1})$ given by (2.4.5) and (2.4.6) is sufficient for $V_t(x_t, u^{t-1}) = x_t$. The stochastic kernels $K_t^{\hat{x}}$ and $G_t^{\hat{x},x}$ are given by

(3.4.6) $$K_t^{\hat{x}}(\hat{x}_t, u_t; \hat{x}_{t+1} \in A) \sim N(\Phi_t \hat{x}_t + \Lambda_t u_t \,,\, \bar{\Sigma}_{t+1})$$

where

(3.4.7) $$\bar{\Sigma}_{t+1} = \Pi_{t+1}\Psi'_{t+1}R^{-1}_{t+1}[\Psi_{t+1}\Phi_t\Pi_t\Phi'_t\Psi'_t + R_{t+1} + \Psi_{t+1}\Gamma_{t+1}\Psi'_{t+1}]R^{-1}_{t+1}\Psi_{t+1}\Pi_{t+1}$$

and

(3.4.8) $$G_t^{\hat{x},x}(\hat{x}_t, x_t \in A) \sim N(\hat{x}_t, \Pi_t) \quad .$$

Proof: It will be shown that $K_t^{\hat{x}}$ defined by (3.4.6) satisfies (3.2.5). For the linear Gaussian model, from Theorem 2.4.1 K_t is given by (2.4.9). The distribution

$$K_t(z^t, u^t; \{z_{t+1} | \hat{x}_{t+1}(z^{t+1}, u^t) \in A\})$$

is obtained from the transformation

(3.4.9) $$\hat{x}_{t+1} = \Phi_t\hat{x}_t + \Lambda_t u_t + \Pi_{t+1}\Psi'_{t+1}R^{-1}_{t+1}(z_{t+1} - \Psi_{t+1}\Lambda_t u_t - \Psi_{t+1}\Phi_t\hat{x}_t)$$

where $\hat{x}_t(z^t, u^{t-1})$ and u_t are fixed and from (2.4.9)

(3.4.10) $$z_{t+1} \sim N[\Psi_{t+1}(\Lambda_t u_t + \Phi_t\hat{x}_t),\; R_{t+1} + \Psi_{t+1}(\Gamma_{t+1} + \Phi_t\Pi_t\Phi'_t)\Psi'_{t+1}]$$

Since the transformation (3.4.9) is linear in z_{t+1}, $\hat{x}_{t+1}$ is Gaussian. It remains only to compute its mean and covariance from (2.4.9) and (3.4.10)

(3.4.11) $$\text{mean } \hat{x}_{t+1} = \Phi_t\hat{x}_t + \Lambda_t u_t + \Pi_{t+1}\Psi'_{t+1}R^{-1}_{t+1}[\Psi_{t+1}(\Lambda_t u_t + \Phi_t\hat{x}_t) - \Psi_{t+1}\Lambda_t u_t - \Psi_{t+1}\Phi_t\hat{x}_t]$$
$$= \Phi_t\hat{x}_t + \Lambda_t u_t$$

and from (3.4.7)

(3.4.12) $$\text{cov } \hat{x}_{t+1} = \Pi_{t+1}\Psi'_{t+1}R^{-1}_{t+1}[R_{t+1} + \Psi_{t+1}(\Gamma_{t+1} + \Phi_t\Pi_t\Phi'_t)\Psi'_{t+1}]R^{-1}_{t+1}\Psi_{t+1}\Pi_{t+1}$$
$$= \bar{\Sigma}_{t+1} \quad .$$

Thus from (3.4.11) and (3.4.12) ,

$$K_t(z^t,u^t;\{z_{t+1}|\hat{x}_{t+1}(z^{t+1},u^t)\in A\}\sim N(\Phi_t\hat{x}_t+\Lambda_t u_t,\Psi_{t+1})$$

Thus (3.2.5) holds for $K_t^{\hat{x}}$ given by (3.4.6), and it follows from Lemma 3.2.1 that $\hat{x}_t$ is a sufficient statistic. From (2.4.4) of Theorem 2.4.1 it is clear that (3.2.9) holds for $G_t^{\hat{x},x}$ given by (3.4.8), and it follows from Lemma 3.2.2 that $\hat{x}_t$ is sufficient for x_t .

For certain loss functions it will be possible to reduce the sufficient statistic further. Let

$$\Phi_{T,t}=\Phi_{T-1}\Phi_{T-2}\cdots\Phi_t\,,\qquad t=0,1,\ldots,T-1$$
(3.4.13)
$$\Phi_{T,T}=I$$

and for $\hat{x}_t$ given by (2.4.5) and (2.4.6) , define

(3.4.14) $$\hat{a}_t(z^t,u^{t-1})=A_T\Phi_{T,t}\hat{x}_t(z^t,u^{t-1})$$

where A_T is an $r\times n$ matrix. The dimension of $\hat{a}_t$ will be less than that of $\hat{x}_t$ provided $r<n$. Let $\{\overset{o}{u}\}$ be a control law that satisfies

(3.4.15) $$\overset{o}{u}_t(z^t)=\overset{o}{u}_{t+1}(z^{t+1})=\ldots\overset{o}{u}_{T-1}(z^{T-1})=0\;;$$

that is, no control from t onwards. From (1.3.1) of the model, the fact that $w_{t+1},\ldots,w_T$ have mean zero and are independent of z^t , (2.4.4), (3.4.13) and (3.4.14), it follows that

(3.4.16)
$$\begin{aligned}E_{\{\overset{o}{u}\}}[A_Tx_T|z^t]&=E_{\{\overset{o}{u}\}}[A_T(\Phi_{T-1}x_{T-1}+w_T)|z^t]\\&=A_T\Phi_{T-1}E_{\{\overset{o}{u}\}}[\Phi_{T-2}x_{T-2}+w_{T-1}|z^t]\\&\;\;\vdots\\&=A_T\Phi_{T-1}\cdots\Phi_{t+1}E_{\{\overset{o}{u}\}}[\Phi_tx_t+w_{t+1}|z^t]\\&=A_T\Phi_{T,t}E_{\{\overset{o}{u}\}}[x_t|z^t]\\&=A_T\Phi_{T,t}\hat{x}_t(z^t,u^{t-1}(z^{t-1}))\\&=\hat{a}_t(z^t,u^{t-1}(z^{t-1}))\qquad\text{a.s. }P_{\{\overset{o}{u}\}}\,.\end{aligned}$$

Thus if $A_T x_T$ represents a final miss vector, then $\hat{a}_t$ is the expected miss vector conditioned on the data z^t available at time t and assuming that no control is exercised from time t on.

Theorem 3.4.3. The statistic $\hat{a}_t(z^t, u^{t-1})$ defined by (3.4.14) is sufficient for

$$(3.4.17) \qquad a_t(x^t, u^{t-1}) = A_T \Phi_{T,t} x_t$$

and the control distributions $K_t^{\hat{a}}$ and $G_t^{\hat{a},a}$ are given by

$$(3.4.18) \qquad K_t^{\hat{a}}(\hat{a}_t, u_t; \hat{a}_{t+1} \in A) \sim N(\hat{a}_t + \zeta_t u_t, \bar{\bar{\Sigma}}_{t+1})$$

where

$$(3.4.19) \qquad \zeta_t = A_T \Phi_{T,t+1} \Lambda_t$$

$$(3.4.20) \qquad \bar{\bar{\Sigma}}_{t+1} = A_T \Phi_{T,t+1} \Pi_{t+1} \Psi'_{t+1} R_{t+1}^{-1} [\Psi_{t+1} \Phi_t \Pi_t \Phi'_t \Psi'_{t+1} + \Psi_{t+1} \Gamma_{t+1} \Psi'_{t+1} + R_{t+1}] R_{t+1}^{-1} \Psi_{t+1} \Pi_{t+1} \Phi'_{T,t+1} A'_T$$

and

$$(3.4.21) \qquad G_t^{\hat{a},a}(\hat{a}_t; a_t \in A) \sim N(\hat{a}_t, A_T \Phi_{T,t} \Pi_t \Phi'_{T,t} A'_T) \quad .$$

Proof: As in the previous theorem, it will be shown that $K_t^{\hat{a}}$ defined by (3.4.18) satisfies (3.2.5) of Lemma 3.2.1. Here from (3.4.14), (2.4.5) of Theorem 2.4.1, (3.4.13), and (3.4.19) $\hat{a}_t$ satisfies

$$(3.4.22) \qquad \hat{a}_{t+1} = \hat{a}_t + \zeta_t u_t + A_T \Phi_{T,t+1} \Pi_{t+1} \Psi'_{t+1} R_{t+1}^{-1} (z_{t+1} - \Psi_{t+1} \Lambda_t u_t - \Psi_{t+1} \Phi_t \hat{x}_t)$$

As before, from (2.4.9), z_{t+1} has normal distribution (3.4.10) where $\hat{x}_t$ and u_t are fixed. Thus $\hat{a}_{t+1}$ is Gaussian with

$$\text{mean } \hat{a}_{t+1} = \hat{a}_t + \zeta_t u_t$$

and

$$\text{cov } \hat{a}_{t+1} = A_T \Phi_{T,t} \Pi_{t+1} \Psi'_{t+1} R_{t+1}^{-1} [\text{cov } z_{t+1}] R_{t+1}^{-1} \Psi_{t+1} \Pi_{t+1} \Phi'_{T,t} A'_T = \bar{\bar{\Sigma}}_{t+1} \, .$$

Thus

$$K_t(z^t, u^t; \{z_{t+1} | \hat{a}_{t+1}(z^{t+1}, u^t) \in A\}) \sim N(\hat{a}_t + \zeta_t u_t, \bar{\bar{\Sigma}}_{t+1})$$

and (3.2.5) holds for $\hat{a}_{t+1}$ and $K_t^{\hat{a}}$ defined by (3.4.18). It follows from Lemma 3.2.1 that $\hat{a}_t$ is a sufficient statistic. From 2.4.4

$$G_t(z^t, u^{t-1}; A_T\Phi_{T,t}x_t \in A) \sim N(A_T\Phi_{T,t}\hat{x}_t, A_T\Phi_{T,t}\Pi_t\Phi'_{T,t}A'_T) \quad .$$

Thus (3.2.9) holds for $a_t = V_t$ given by (3.4.17) and $G_t^{\hat{a},a}$ by (3.4.21). Thus from Lemma 3.2.2 $\hat{a}_t$ is sufficient for a_t .

<u>Theorem</u> 3.4.4 i) For a linear Gaussian system with loss function of the form

$$(3.4.23) \qquad L = \sum_{t=1}^{T} c_t^V(x_t \, , \sum_{\tau=0}^{t-1} \bar{c}_\tau(u_\tau), u_{t-1})$$

where the $\bar{c}_\tau$ are measurable functions taking values in a finite dimensional Euclidean space, the statistic

$$(3.4.24) \qquad Y_t = (\hat{x}_t \, , \sum_{\tau=0}^{t-1} \bar{c}_\tau(u_\tau))$$

is sufficient.

ii) For the loss function

$$(3.4.25) \qquad L = \sum_{t=1}^{T} c_t^V(A_T\Phi_{T,t}x_t, \sum_{\tau=0}^{t-1} \bar{c}_\tau(u_\tau), u_{t-1}) \quad ,$$

the statistic

$$(3.4.26) \qquad Y_t = (\hat{a}_t, \sum_{\tau=0}^{t-1} \bar{c}_\tau(u_\tau))$$

is sufficient.

Proof: These results follow immediately from Theorems 3.4.2 and 3.4.3 and Corollary 3.3.1.

<u>Theorem</u> 3.4.5 i) If the matrices $\bar{\Phi}_t$ given by (3.4.7) are non-singular for $t = 1,2,\ldots$ then the statistic $\hat{x}_t$ is dominated by

$$(3.4.27) \qquad \mu^{(n)} = \{\mu^n, \mu^n, \ldots\}$$

where μ^n is n-dimensional Lebesgue measure.

ii) If the $\bar{\bar{\Phi}}_t$ given by (3.4.20) are non-singular, then the statistic $\hat{a}_t$ is dominated by

$$(3.4.28) \qquad \mu^{(r)} = \{\mu^r, \mu^r, \ldots\}$$

where μ^r is r-dimensional Lebesgue measure.

Proof: This follows from Theorem 3.4.2 (Theorem 3.4.3) and Lemma 3.2.5.

Chapter 4 - General Theory of Optimality

4.1 Introduction

In this chapter the only requirements made of the loss function $L(x^T, z^T, u^{T-1})$ are that it be measurable, non-negative, and possibly infinite. The principal idea exploited here is that of the minimal conditional loss functional $\hat{F}_t(z^t, \{u\})$. For $\{u\}$ fixed $\hat{F}_t$ is a measurable function of z^t, and for z^t fixed it depends on the truncated law $\{u^{t-1}\}$ - hence the term "functional." In the usual treatment of the dynamic programming problem (Blackwell [2], Hinderer [8], etc) the minimal conditional loss function depends only on the value of the truncated law $u^{t-1}(z^{t-1})$ at the observation point z^t and not on the entire law $\{u^{t-1}\}$. While the complexity of the functional $\hat{F}_t(z^t, \{u\})$ over the function $\hat{\mathfrak{F}}_t(z^t, u^{t-1})$ may be a disadvantage, it appears to be outweighed by the "natural" properties of $\hat{F}_t$. Section 4.4, in which a comparison of the two criteria is made is not required for the development of the argument and is included primarily to show the relation of the present approach to that of the literature of dynamic programming.

The principal advantage of the functional over the function $\mathfrak{F}_t(z^t, u^{t-1})$ is that the functional $\hat{F}_t(z^t, \{u\})$ exists and is Borel measurable in z^t for arbitrary observation space Z_t, control space $\mathcal{U}^t$, and class $\mathcal{D}$ of admissible control laws. The minimal loss function $\hat{\mathfrak{F}}_t(z^t, u^{t-1})$ is at best universally measurable, and even this requires the assumption of additional (though natural) properties of the spaces Z^t and $\mathcal{U}^{t-1}$. More serious is the requirement that the class of admissible laws $\mathcal{D}$ contain essentially all measurable laws. (See Strauch [16] Theorem 7.1 and Hinderer [8] Theorem 1.3.2). In section 4.4 assumptions are given under which the two criteria are the same in the sense that

$$\hat{F}_t(z^t, \{u\}) = \hat{\mathfrak{F}}_t(z^t, u^{t-1}(z^{t-1})) \qquad \text{a.s. } P^t_{\{u\}}$$

for all $\{u\}$, and an example is given to show that this equality does not always hold.

In section 4.2 properties of the functionals $F_t(z^t,\{u\})$ are defined - "non-negative", "compatible", "martingale" , "sub-martingale", and "closed" . The conditional loss functional $\hat{F}_t$ and the step-wise conditional loss functional $\tilde{F}_t$ are defined making use of the idea of the P-ess inf of a class of non-negative, measurable functions. The definition and required properties of the P-ess inf are given in the appendix section A.2. Also defined in section 4.2 are the finite and countable ϵ-lattice properties of a control system. In the remainder of the section, martingale properties of and relationships between the conditional loss functionals $\hat{F}_t$ and $\tilde{F}_t$ are developed. Theorems 4.2.4 and 4.2.5 are the analogues for countably additive measures of Theorem 1 and Corollary 2 of section 2.14 of Dubins and Savage [5] . The relationship between excessive functions and super-martingales is discussed in section 2.12 of [5].

Section 4.3 is devoted to necessary and sufficient conditions for optimality at a control law. The motivation for the definition of optimal (ϵ-optimal) for $\{u\}$ at t (Definitions 4.3.1 and 4.3.2) is as follows: control law $\{u\}$ has for whatever reason been used and data has been collected up through time t . At this point the question is asked how best to proceed from t onward. The property optimal (ϵ-optimal) for $\{u\}$ at t is the same as p-optimal ((p,ϵ)-optimal) of Blackwell [2] where $t = 1$ and $P^t_{\{u\}} = p$. This property is given in Theorem 4.3.1 and could as well be used as the defining property. The defining properties given in Definition 4.3.1 and 4.3.2 were selected because they are more constructive and more convenient in applications. The properties given in Theorem 4.3.2 are the same as $\bar{p}$-optimal and $\bar{p}$-ϵ-optimal of Hinderer [8] again with $t = 1$ and $P^t_{\{u\}} = p$.

Under the assumption that the system has the countable ϵ-lattice property, it is shown in Theorem 4.3.4 that ϵ-optimal controls always exist. For systems which do not enjoy this property the weaker property (4.3.13) of Theorem 4.3.2 ($\bar{p}$-ϵ-optimum) is probably a more basic and a better defining property for ϵ-optimality. In fact all the results of this chapter are of practical interest only for systems with the countable ϵ-lattice property.

In Theorems 4.3.5, 4.3.6, and 4.3.7 a step-wise method of computing ϵ-optimal controls is given which is very similar to the standard dynamic programming method. The minimum loss function V_t is replaced by $\hat{F}_t$, and minimization in the control value u_t is replaced by the $P^t_{\{u\}}$-ess inf in the control function $u_t(\cdot)$.

The remainder of section 4.3 is devoted to relationships between optimality and the martingale properties of $\hat{F}_t$. The most easily applicable sufficient condition for optimality is given in Lemma 4.3.3. In Theorem 4.3.8 under additional assumptions necessary and sufficient conditions for optimality of a control law are provided.

In section 4.4 a stronger type of optimality (in the sense of Strauch) is defined. Theorem 4.4.1 justifies restricting ones attention to controls optimal in the sense defined in section 4.3 under the following reasoning. Suppose an ϵ-optimal control $\{\hat{u}^\epsilon\}$ has been found and suppose controls ϵ-optimal in the sense of Strauch exist for all $\epsilon > 0$, then Theorem 4.4.1 assures that the control $\{\hat{u}^\epsilon\}$ also has the stronger optimality property, in the sense of Strauch.

4.2 A Conditional loss functional

In this section an arbitrary non-negative possibly infinite loss function L is considered, and $\mathcal{D}$ denotes a class of control laws with possibly infinite expected loss.

Definition 4.2.1. A functional $F_t(z^t,\{u\})$ is a non-negative, compatible functional provided for $t = 0,1,\ldots,T$.

i) $F_t(z^t,\{u\})$ is $\mathcal{B}^t_Z$-measurable in z^t for all $\{u\} \in \mathcal{D}$;

ii) for $\{\bar{u}\} \in \mathcal{D}$, $\{u\} \in \mathcal{D}$ such that $\{\bar{u}^{t-1}\} = \{u^{t-1}\}$,

$F_t(z^t,\{u\}) = F_t(z^t,\{\bar{u}\})$;

and

iii) $0 \leq F_t(z^t,\{u\}) \leq \infty$ a.s. $P^t_{\{u\}}$

for all $\{u\} \in \mathcal{D}$.

By convention consistent with i), ii), and (1.2.8) F_0 is a non-negative real number independent of $\{u\} \in \mathscr{D}$.

The effect of condition ii) is that $F_t(z^t,\{u\})$ does not depend on the entire control law $\{u\}$ but only on the truncated law $\{u^{t-1}\}$. A compatible functional will occasionally be written $F_t(z^t,\{u^{t-1}\})$ in order to emphasis this fact.

Compatible functionals which are also submartingales or martingales will be of particular interest. The term "closed" in this connection will be used to mean closed in the sense of martingales by $F_T = E_{\{u\}}[L|z^T]$.

Definition 4.2.2. The non-negative, compatible functional $F_\tau(z^\tau,\{u\})$ is a (sub-) martingale for $\{u\}$ and $t \leq \tau \leq T$ provided

$$\text{i)} \quad F_\tau(z^\tau,\{u\}) \ (\leq) \ E_{\{u\}}[F_{\tau+1}(z^{\tau+1},\{u\})|z^\tau] \qquad \text{a.s. } P^\tau_{\{u\}}$$

for $\tau = t\ , t+1,\ldots,T-1$. It is a closed (sub-) martingale for $\{u\}$ if in addition

$$\text{ii)} \quad F_T(z^T,\{u\}) = E_{\{u\}}[L|z^T] \qquad \text{a.s. } P^T_{\{u\}} \ .$$

The next two definitions make use of the P-ess inf whose definition and properties are given in the appendix section A.2.

Definition 4.2.3. The conditional loss functional $\hat{F}^{\mathscr{D}}$ is defined by

$$(4.2.1) \qquad \hat{F}^{\mathscr{D}}_t(z^t,\{u\}) = P^t_{\{u\}}\text{-ess inf}_{\substack{\{\bar{u}\}\in\mathscr{D},\\ \{\bar{u}^{t-1}\}=\{u^{t-1}\}}} E_{\{\bar{u}\}}[L|z^t]$$

for $t = 1,2,\ldots,T$ and $\{u\} \in \mathscr{D}$. By convention, consistent with (1.2.19) and Lemma A.2.6

$$(4.2.2) \qquad \hat{F}^{\mathscr{D}}_0 = \inf_{\{\bar{u}\}\in\mathscr{D}} E_{\{\bar{u}\}}[L] \ .$$

Definition 4.2.4. The step-wise conditional loss functional $\tilde{F}^{\mathscr{D}}_t$ is defined by backward induction

$$(4.2.3) \qquad \tilde{F}^{\mathscr{D}}_T(z^T,\{u\}) = E_{\{u\}}[L|z^T] \ ,$$

$$(4.2.4) \qquad \tilde{F}_t^{\mathscr{D}}(z^t,\{u\}) = P^t_{\{u\}}\text{-ess}\inf_{\substack{\{\bar{u}\}\in\mathscr{D} \ni \\ \{\bar{u}^{t-1}\}=\{u^{t-1}\}}} E_{\{\bar{u}\}}[\tilde{F}^{\mathscr{D}}_{t+1}(z^{t+1},\{\bar{u}\})|z^t]$$

for $t = T\text{-}1, T\text{-}2,\ldots,1$; $\{u\} \in \mathscr{D}$, and again by convention

$$(4.2.5) \qquad \tilde{F}_0^{\mathscr{D}} = \inf_{\{\bar{u}\}\in\mathscr{D}} E_{\{\bar{u}\}}[\tilde{F}_1^{\mathscr{D}}(z_1,\{\bar{u}\})] \quad .$$

The superscript $\mathscr{D}$ will be omitted when the intended class of control laws is clear. Some immediate properties of these functionals will be stated as a lemma.

Lemma 4.2.1. The functionals $\hat{F}_t(z^t,\{u\})$ and $\tilde{F}_t(z^t,\{u\})$ exist, are defined uniquely a.s. $P^t_{\{u\}}$ by (4.2.1)-(4.2.2) and (4.2.3)-(4.2.5) respectively, are non-negative compatible functinals, and $\hat{F}_T$ satisfies

$$(4.2.6) \qquad \hat{F}_T(z^T,\{u\}) = E^T_{\{u\}}[L|z^T] \qquad \text{a.s. } P^T_{\{u\}}$$

for all $\{u\} \in \mathscr{D}$.

Proof: The existence and a.s. uniqueness of $\hat{F}_t$ and $\tilde{F}_t$ follow from Theorem A.2.1 and Lemma A.2.9., that of $\tilde{F}_t$ follows by backward induction. The measurability requirement i) of Definition 4.2.1 follows from i) of Definition A.2.1; ii) of Definition 4.2.1 is obvious from (4.2.1) and (4.2.4). Since L is non-negative,

$$0 \leq E_{\{\bar{u}\}}[L|z^t] \qquad \text{a.s. } P^t_{\{u\}}$$

for all $\{\bar{u}\} \in \mathscr{D}$. Thus iii) for $\hat{F}_t$ follows from (4.2.1) and iii) of Definition A.2.1. For $\tilde{F}_t$, iii) follows by backward induction from (4.2.4) and iii) of Definition A.2.1. Since the distribution of L is determined by $\{u^{T-1}\} = \{u\}$, the family over which the P-ess inf is taken in (4.2.1) for $t = T$ contains the single member $\{u\}$. The result (4.2.6) then follows from Lemma A.2.7.

Lemma 4.2.2. For all $\{u\} \in \mathcal{D}$, $\tilde{F}_t(z^t,\{u\})$ is a closed submartingale.

Proof: From (4.2.4) and ii) of Definition A.2.1

$$\tilde{F}_t(z^t,\{u\}) \leq E_{\{u\}}[\tilde{F}_{t+1}(z^{t+1},\{u\})|z^t] \qquad \text{a.s. } P^t_{\{u\}}$$

Lemma 4.2.3. For all $\{u\} \in \mathcal{D}$ and $0 \leq t \leq T$

$$(4.2.7) \qquad \tilde{F}_t(z^t,\{u\}) \leq P^t_{\{u\}}\text{-ess inf}_{\substack{\{\bar{u}\} \in \mathcal{D} \ni \\ \{\bar{u}^{t-1}\} = \{u^{t-1}\}}} E_{\{\bar{u}\}}[\hat{F}_{t+1}(z^{t+1},\{\bar{u}\})|z^t]$$

$$\leq \hat{F}_t(z^t,\{\bar{\bar{u}}\}) \qquad \text{a.s. } P^t_{\{u\}} \quad .$$

Proof: From ii) of Definition 4.2.1, the submartingale property of $\tilde{F}_t(z^t,\bar{\bar{u}})$ and (4.2.3) of Definition 4.2.4, for $\{\bar{\bar{u}}\} \in \mathcal{D}$ such that $\{\bar{\bar{u}}^t\} = \{\bar{u}^t\}$

$$\tilde{F}_{t+1}(z^{t+1},\{\bar{u}\}) = \tilde{F}_{t+1}(z^{t+1},\{\bar{\bar{u}}\})$$

$$\leq E_{\{\bar{\bar{u}}\}}[\tilde{F}_T(z^T,\{\bar{\bar{u}}\})|z^{t+1}] = E_{\{\bar{\bar{u}}\}}[L|z^{t+1}] \qquad \text{a.s. } P^{t+1}_{\{\bar{u}\}} \quad .$$

Thus from Definitions 4.2.3 and A.2.1

$$\tilde{F}_{t+1}(z^{t+1},\{\bar{u}\}) \leq \hat{F}_{t+1}(z^{t+1},\{\bar{u}\}) \leq E_{\{\bar{\bar{u}}\}}[L|z^{t+1}] \qquad \text{a.s. } P^{t+1}_{\{\bar{u}\}}$$

for all $\{\bar{u}\} \in \mathcal{D}$. Taking conditional expectations

$$E_{\{\bar{u}\}}[\tilde{F}_{t+1}(z^{t+1},\{\bar{u}\})|z^t] \leq E_{\{\bar{u}\}}[\hat{F}_{t+1}(z^{t+1},\{\bar{u}\})|z^t]$$

$$\leq E_{\{\bar{u}\}}[L|z^t] \qquad \text{a.s. } P^t_{\{\bar{u}\}} \quad .$$

Then from (4.2.4), Lemma A.2.8, and (4.2.1)

$$\tilde{F}_t(z^t,\{u\}) = P^t_{\{u\}}\text{-ess inf}_{\substack{\{\bar{u}\} \in \mathcal{D} \ni \\ \{\bar{u}^{t-1}\} = \{u^{t-1}\}}} E_{\{\bar{u}\}}[\tilde{F}_{t+1}(z^{t+1},\{\bar{u}\})|z^t]$$

$$\leq P^t_{\{u\}}\text{-ess inf}_{\substack{\{\bar{u}\} \in \mathcal{D} \ni \\ \{\bar{u}^{t-1}\} = \{u^{t-1}\}}} E_{\{\bar{u}\}}[\hat{F}_{t+1}(z^{t+1},\{\bar{u}\})|z^t]$$

$$\leq P^t_{\{u\}}\text{-ess inf}_{\substack{\{\bar{u}\} \in \mathcal{D} \ni \\ \{\bar{u}^{t-1}\} = \{u^{t-1}\}}} E_{\{\bar{u}\}}[L|z^t] = \hat{F}_t(z^t,\{u\}) \qquad \text{a.s. } P^t_{\{u\}} \quad .$$

Lemma 4.2.4. If the non-negative compatible functional $F_\tau(z^\tau,\{u\})$ is a closed submartingale for $t \leq \tau \leq T$ and all $\{u\} \in \mathcal{D}$, then

$$(4.2.8) \qquad F_\tau(z^\tau,\{u\}) \leq \tilde{F}_\tau(z^\tau,\{u\}) \qquad \text{a.s. } P^T_{\{u\}}$$

for all $\{u\} \in \mathcal{D}$ and $t \leq \tau \leq T$.

Proof: The proof is by backward induction. For $\tau = T$, (4.2.8) follows from ii) of Definition 4.2.2 and (4.2.3). Assume then that (4.2.8) holds for $\tau + 1$. For $\{\bar{u}\}$ such that $\{\bar{u}^{\tau-1}\} = \{u^{\tau-1}\}$, from ii) of Definition 4.2.1, the submartingale property of $F_\tau(z^\tau,\{\bar{u}\})$, and the induction hypothesis

$$F_\tau(z^\tau,\{u\}) = F_\tau(z^\tau,\{\bar{u}\}) \leq E_{\{\bar{u}\}}[F_{\tau+1}(z^{\tau+1},\{\bar{u}\})|z^\tau]$$

$$\leq E_{\{\bar{u}\}}[\tilde{F}_{\tau+1}(z^{\tau+1},\{\bar{u}\}|z^\tau] \qquad \text{a.s. } P^T_{\{u\}} .$$

Thus from iii) of Definition A.2.1 and Definition 4.2.4

$$F_\tau(z^\tau,\{u\}) \leq P^T_{\{u\}}\text{-ess}\inf_{\substack{\{\bar{u}\} \in \mathcal{D} \ni \\ \{\bar{u}^{\tau-1}\} = \{u^{\tau-1}\}}} E_{\{\bar{u}\}}[\tilde{F}_{\tau+1}(z^{\tau+1},\{u\})|z^\tau]$$

$$= \tilde{F}_\tau(z^\tau,\{u\}) \qquad \text{a.s. } P^T_{\{u\}} .$$

Theorem 4.2.1. The step-wise conditional loss function $\tilde{F}_t$ is characterized by the property that it is the maximal non-negative, compatible, closed submartingale. That is,

i) $\tilde{F}_t(z^t,\{u\})$ is a non-negative,compatible, closed submartingale for $0 \leq t \leq T$ and all $\{u\} \in \mathcal{D}$; and

ii) if $F_\tau(z^\tau,\{u\})$ is a non-negative, compatible, closed submartingale for $t \leq \tau \leq T$ and all $\{u\} \in \mathcal{D}$, then

$$F_\tau(z^\tau,\{u\}) \leq \tilde{F}_\tau(z^\tau,\{u\}) \qquad \text{a.s. } P^T_{\{u\}}$$

for all $\{u\} \in \mathcal{D}$ and $t \leq \tau \leq T$.

Proof: It follows from Lemmas 4.2.1, 4.2.2, and 4.2.4 that the step-wise conditional loss functional $\tilde{F}_t$ satisfies i) and ii). Let $\tilde{\tilde{F}}_t$ satisfy i) and ii). From i) and Lemma 4.2.4 with $t = 0$, for $0 \leq \tau \leq T$

$$\tilde{\tilde{F}}_\tau(z^\tau,\{u\}) \leq \tilde{F}_\tau(z^\tau,\{u\}) \qquad \text{a.s. } P^\tau_{\{u\}} \ .$$

Since $\tilde{F}_\tau$ is a non-negative, compatible, closed submartingale for $0 \leq \tau \leq T$, from ii)

$$\tilde{F}_\tau(z^\tau,\{u\}) \leq \tilde{\tilde{F}}_\tau(z^\tau,\{u\}) \qquad \text{a.s. } P^\tau_{\{u\}} \ .$$

Thus

$$\tilde{F}_\tau(z^\tau,\{u\}) = \tilde{\tilde{F}}_\tau(z^\tau,\{u\}) \qquad \text{a.s. } P^\tau_{\{u\}}$$

for $0 \leq \tau \leq T$ and all $\{u\} \in \mathfrak{D}$.

Theorem 4.2.2. The conditional loss functional $\hat{F}_t(z^t,\{u\})$ is a submartingale for $0 \leq t \leq T$ and all $\{u\} \in \mathfrak{D}$ if and only if

$$\hat{F}_t(z^t,\{u\}) = \tilde{F}_t(z^t,\{u\}) \qquad \text{a.s. } P^t_{\{u\}} \tag{4.2.9}$$

for $0 \leq t \leq T$ and all $\{u\} \in \mathfrak{D}$.

Proof: If $\hat{F}_t$ is a submartingale, then from Lemma 4.2.1 it is a non-negative, compatible, closed submartingale. Thus from Theorem 4.2.1

$$\hat{F}_t(z^t,\{u\}) \leq \tilde{F}_t(z^t,\{u\}) \qquad \text{a.s. } P^t_{\{u\}} \ .$$

The result (4.2.9) then follows from Lemma 4.2.3. If (4.2.9) holds, then from i) of Theorem 4.2.1 $\hat{F}_t$ is a submartingale.

Conditions on the control system will be introduced next under which the conditional loss function is a submartingale. The ε-lattice properties are essentially completeness or robustness properties of the class of admissible controls $\mathfrak{D}$.

Definition 4.2.5. The control system has the finite ε-lattice property provided for $\{\bar{u}\} \in \mathfrak{D}$ and $\{\bar{\bar{u}}\} \in \mathfrak{D}$ such that

$$\{\bar{u}^{t-1}\} = \{\bar{\bar{u}}^{t-1}\} \tag{4.2.10}$$

and $\varepsilon > 0$, these exists $\{\tilde{u}\} \in \mathcal{D}$ which satisfies

$$(4.2.11) \qquad \{\tilde{u}^{t-1}\} = \{\bar{u}^{t-1}\}$$

and

$$(4.2.12) \qquad E_{\{\tilde{u}\}}[L|z^t] \leq E_{\{\bar{u}\}}[L|z^t] + \varepsilon \qquad \text{a.s. } P^t_{\{\tilde{u}\}}$$

$$(4.2.13) \qquad E_{\{\tilde{u}\}}[L|z^t] \leq E_{\{\bar{\bar{u}}\}}[L|z^t] + \varepsilon \qquad \text{a.s. } P^t_{\{\tilde{u}\}} \quad .$$

Definition 4.2.6. The control system has the countable ε-lattice property provided for all t , $\varepsilon > 0$, and countable sub-class $\bar{\mathcal{D}} \subset \mathcal{D}$ which satisfies (4.2.10) for all $\{\bar{u}\}$, $\{\bar{\bar{u}}\} \in \bar{\mathcal{D}}$, there exists $\{\tilde{u}\} \in \mathcal{D}$ which satisfies (4.2.11) and (4.2.12) for all $\{\bar{u}\} \in \bar{\mathcal{D}}$.

In Definitions A.2.2 and A.2.3 of the appendix, the ε-lattice properties are attributed to families of measurable, a.s. non-negative functions. For t and $\{u\} \in \mathcal{D}$ fixed such a family of functions is provided by

$$(4.2.14) \qquad \{E_{\{\bar{u}\}}[L|z^t]\}_{\{\bar{u}\} \in \mathcal{D} \ni \{\bar{u}^{t-1}\} = \{u^{t-1}\}} \quad .$$

The lattice property holds for a control system provided it holds for the family of functions (4.2.14) for all t and $\{u\} \in \mathcal{D}$.

Theorem 4.2.3. If the control system has the finite ε-lattice, then $\hat{F}_t(z^t,\{u\})$ is a submartingale for all $\{u\}$ which satisfy

$$(4.2.15) \qquad E_{\{u\}}[L] < \infty \quad .$$

Proof: For t fixed and $\{u\}$ satisfying (4.2.15) , (4.2.14) is a family of measurable, a.s. non-negative functions with the finite ε-lattice property. Further, it contains the integrable member

$$E_{\{u\}}[L|z^t] \; .$$

Thus from the definition of $\hat{F}_{t+1}$, Theorem A.2.2 on $(Z^{t+1}, \mathcal{B}_Z^{t+1}, P_{\{u\}})$ with $\mathcal{B} = \mathcal{B}_Z^t$, Lemma A.2.10, and the definition of $\hat{F}_t$

$$E_{\{u\}}[\hat{F}_{t+1}(z^{t+1},\{u\})|z^t]$$

$$= E_{\{u\}}\{P^{t+1}_{\{u\}}\text{-ess}\inf_{\substack{\{\bar{u}\} \in \mathcal{D} \ni \\ \{\bar{u}^t\} = \{u^t\}}} E_{\{\bar{u}\}}[L|z^{t+1}]|z^t\}$$

$$= P^t_{\{u\}}\text{-ess}\inf_{\substack{\bar{u} \in \mathcal{D} \ni \\ \{\bar{u}^t\} = \{u^t\}}} E_{\{\bar{u}\}}[E_{\{\bar{u}\}}[L|z^{t+1}]|z^t]$$

$$= P^t_{\{u\}}\text{-ess}\inf_{\substack{\{\bar{u}\} \in \mathcal{D} \ni \\ \{\bar{u}^t\} = \{u^t\}}} E_{\{\bar{u}\}}[L|z^t]$$

$$\geq P^t_{\{u\}}\text{-ess}\inf_{\substack{\{\bar{u}\} \in \mathcal{D} \ni \\ \{\bar{u}^{t-1}\} = \{u^{t-1}\}}} E_{\{\bar{u}\}}[L|z^t] = \hat{F}_t(z^t,\{u\}) \quad \text{a.s. } P^t_{\{u\}} \quad .$$

Thus $\hat{F}_t(z^t,\{u\})$ is a submartingale.

Theorem 4.2.4. If the control system has the countable ϵ-lattice property, then $\hat{F}_t(z^t,\{u\})$ is a submartingale for all $\{u\} \in \mathcal{D}$ and

(4.2.16) $$\hat{F}_t(z^t,\{u\}) = \tilde{F}_t(z^t,\{u\}) \qquad \text{a.s. } P^t_{\{u\}}$$

for all $\{u\} \in \mathcal{D}$ and $0 \leq t \leq T$.

Proof: The proof of the submartingale property is the same as that of the previous theorem with Theorem A.2.3 used in place of Theorem A.2.2. The result (4.2.16) follows from Theorem 4.2.2.

Theorem 4.2.5. If the control system has the countable ϵ-lattice property, then $\hat{F}_t(z^t,\{u\})$ is characterized by the properties i) and ii) of Theorem 4.2.1.

Proof: This follows from Theorem 4.2.1 and (4.2.16) of Theorem 4.2.4.

4.3 Optimality for $\{u\}$ at time t

Two types of optimality of a control law will be defined. The intent of the definitions is made clear by the properties given in Theorems 4.3.1 and 4.3.2. The heuristic motivative is discussed in section 4.1.

Definition 4.3.1. Let $\{u\} \in \mathcal{D}$ and $1 \leq t \leq T-1$. Then $\{\hat{u}\}$ is optimal for $\{u\}$ at t provided

$$(4.3.1) \qquad \{\hat{u}\} \in \mathcal{D} \quad , \quad \{\hat{u}^{t-1}\} = \{u^{t-1}\} \quad ,$$

and

$$(4.3.2) \qquad E_{\{\hat{u}\}}[L|z^t] = \hat{F}_t(z^t,\{u\}) \qquad \text{a.s. } P^t_{\{u\}} \quad .$$

The law $\{\hat{u}\} \in \mathcal{D}$ is optimal at $t = 0$ provided

$$(4.3.3) \qquad E_{\{\hat{u}\}}[L] = \hat{F}_0 \quad .$$

Definition 4.3.2. Let $\{u\} \in \mathcal{D}$, $1 \leq t \leq T-1$, and $\epsilon > 0$. Then $\{\hat{u}^\epsilon\}$ is ϵ-optimal for $\{u\}$ at t provided

$$(4.3.4) \qquad \{\hat{u}^\epsilon\} \in \mathcal{D} \quad , \quad \{\hat{u}^{t-1,\epsilon}\} = \{u^{t-1}\} \quad ,$$

and

$$(4.3.5) \qquad E_{\{\hat{u}^\epsilon\}}[L|z^t] \leq \hat{F}_t(z^t,\{u\}) + \epsilon \qquad \text{a.s. } P^t_{\{u\}} \quad .$$

The law $\{\hat{u}^\epsilon\} \in \mathcal{D}$ is ϵ-optimal at $t = 0$ provided

$$(4.3.6) \qquad E_{\{\hat{u}^\epsilon\}}[L] \leq \hat{F}_0 + \epsilon \quad .$$

The notation $\{\hat{u}^\epsilon\}$ indicates a superscript ϵ on the control functions $u^\epsilon_t(z^t)$ and is not to be confused with the notation $\{u^t\}$ for a truncated control law. Trucation of $\{\hat{u}^\epsilon\}$ is indicated as in (4.3.4) by $\{\hat{u}^{t-1,\epsilon}\}$. The superscript ϵ will be omitted when the intention is clear.

The Definitions 4.3.1 and 4.3.2 are a bit redundant in that 4.3.2 can be obtained from 4.3.1 by letting $\epsilon = 0$. It is, however, convenient to have two different definitions since in most instances the properties of and conditions for the two types of optimality are quite different. Whenever

ϵ occurs, it will always be assumed that $\epsilon > 0$.

<u>Theorem</u> 4.3.1 i) The control law $\{\hat{u}\}$ is optimal for $\{u\}$ if and only if $\{\hat{u}\}$ satisfies (4.3.1) and

$$E_{\{\hat{u}\}}[L|z^t] \leq E_{\{\bar{u}\}}[L|z^t] \qquad \text{a.s. } P^t_{\{u\}} \tag{4.3.7}$$

for all $\{\bar{u}\}$ which satisfy

$$\{\bar{u}\} \in \mathcal{D} \quad \text{and} \quad \{\bar{u}^{t-1}\} = \{u^{t-1}\} . \tag{4.3.8}$$

ii) The control law $\{\hat{u}^\epsilon\}$ is ϵ-optimal for $\{u\}$ at t if and only if $\{\hat{u}^\epsilon\}$ satisfies (4.3.4) and

$$E_{\{\hat{u}^\epsilon\}}[L|z^t] \leq E_{\{\bar{u}\}}[L|z^t] + \epsilon \qquad \text{a.s. } P^t_{\{u\}} \tag{4.3.9}$$

for all $\{\bar{u}\}$ which satisfy (4.3.8).

iii) $\{\hat{u}\}$ is obtimal at $t = 0$ if and only if $\{\hat{u}\} \in \mathcal{D}$ and

$$E_{\{\hat{u}\}}[L] \leq E_{\{u\}}[L] \tag{4.3.10}$$

for all $\{u\} \in \mathcal{D}$.

iv) $\{\hat{u}^\epsilon\}$ is ϵ-optimal at $t = 0$ if and only if $\{\hat{u}^\epsilon\} \in \mathcal{D}$ and

$$E_{\{\hat{u}^\epsilon\}}[L] \leq E_{\{u\}}[L] + \epsilon \tag{4.3.11}$$

for all $\{u\} \in \mathcal{D}$.

v) For all $\epsilon > 0$, there exists $\{\hat{u}^\epsilon\}$, ϵ-optimal at $t = 0$.

Proof: The results i)-iv) follow immediately from Definitions 4.3.1 , 4.3.2 , 4.2.3 , and A.2.1 . For $\epsilon > 0$, an ϵ-optimal control at $t = 0$ is obtained by taking $\{\hat{u}^\epsilon\}$ which satisfies

$$E_{\{\hat{u}^\epsilon\}}[L] \leq \inf_{\{u\} \in \mathcal{D}} E_{\{u\}}[L] + \epsilon \quad .$$

From Theorem 4.3.1 ii) it is clear that the existence of ϵ-optimal controls all $\{u\}$, t , and $\epsilon > 0$ is essentially an extension of the ϵ-lattice properties. A system for which ϵ-optimal controls exist obviously has the countable ε-lattice property and might best described as having the universal ε-lattice property.

Theorem 4.3.2. i) If $\{\hat{u}\}$ is optimal for $\{u\}$ at t, then

$$(4.3.12) \qquad E_{\{\hat{u}\}}[L] = \inf_{\substack{\{\bar{u}\} \in \mathscr{D} \ni \\ \{\bar{u}^{t-1}\} = \{u^{t-1}\}}} E_{\{\bar{u}\}}[L]$$

ii) If $\{\hat{u}^{\varepsilon}\}$ is ε-optimal for $\{u\}$ at t, then

$$(4.3.13) \qquad E_{\{\hat{u}^{\varepsilon}\}}[L] \leq \inf_{\substack{\{\bar{u}\} \in \mathscr{D} \ni \\ \{\bar{u}^{t-1}\} = \{u^{t-1}\}}} E_{\{\bar{u}\}}[L] + \varepsilon$$

Proof: Let $\{\hat{u}\}$ be optimal for $\{u\}$ at t. Integrating (4.3.7) of Theorem 4.3.1 i) with respect to $P^t_{\{u\}}$,

$$E_{\{\hat{u}\}}[L|z^t] \leq E_{\{\bar{u}\}}[L|z^t] \qquad \text{a.s. } P^t_{\{u\}}$$

for all $\{\bar{u}\}$ which satisfy (4.3.8),

$$(4.3.14) \qquad E_{\{\hat{u}\}}[L] \leq E_{\{\bar{u}\}}[L] .$$

Thus from (4.3.14) and (4.3.1)

$$E_{\{\hat{u}\}}[L] \leq \inf_{\substack{\{\bar{u}\} \in \mathscr{D} \ni \\ \{\bar{u}^{t-1}\} = \{u^{t-1}\}}} E_{\{\bar{u}\}}[L] \leq E_{\{\hat{u}\}}[L] ,$$

and (4.3.12) follows. Let $\{\hat{u}^{\varepsilon}\}$ be ε-optimal for $\{u\}$ at t, and let $\{\bar{u}\}$ satisfy (4.3.8). As before, integrating (4.3.9) of Theorem 4.3.1 ii)

$$E_{\{\hat{u}^{\varepsilon}\}}[L] \leq E_{\{\bar{u}\}}[L] + \varepsilon ,$$

and (4.3.13) follows.

The conditions (4.3.12) and (4.3.13) are the conditions $\bar{p}$-optimal and $\bar{p}$-ε-optimal of Hinderer [8] where $t = 1$ and $p = P^t_{\{u\}}$.

Under the finite ε-lattice assumption, it will be shown that the converse of Theorem 4.3.2 i) is true for controls laws $\{u\}$ with finite expected loss. The converse of ii) is not true generally. Thus the property (4.3.5) of Definition 4.3.2 is genuinely stronger than (4.3.13).

The following result is given by Hinderer [8] Theorem 20.1 under the assumption that the class $\mathcal{D}$ is stable. The definition of stability and its relationship to the finite ϵ-lattice property are given in Definition 4.4.3 and Theorem 4.4.2 of the next section.

Theorem 4.3.3. Let the control system have the finite ϵ-lattice property and let $\{u\} \in \mathcal{D}$ satisfy

(4.3.15) $$E_{\{u\}}[L] < \infty \quad .$$

Then $\{\hat{u}\}$ is **optimal** for $\{u\}$ at t if and only if it satisfies (4.3.1) and (4.3.12).

Proof: This follows immediately from Definitions 4.3.1, 4.2.3 and Lemma A.2.17.

The existence of ϵ-optimal controls is proved by Blackwell [2] Theorem 1 and Hinderer [8] Theorem 20.3 b) under the assumption of countable stability (see Definition 4.4.4 and Theorem 4.4.2).

Theorem 4.3.4. Let the control system have the countable ϵ-lattice property. Then for each $\{u\} \in \mathcal{D}$, $0 \leq t \leq T-1$ and $\epsilon > 0$, there exists a control law $\{\hat{u}^{\epsilon}\}$ that is ϵ-optimal for $\{u\}$ at t .

Proof: This follows from Lemma A.2.16 and Definitions 4.3.2 and 4.2.3.

In the next series of results it will be shown that under the countable ϵ-lattice assumption, optimal control laws can be computed step-wise in a way very similar to the standard dynamic programming method. The step-wise computation of the conditional loss function is given by the equation (4.2.4). In order to emphasize the step-wise nature of the procedure some additional notation will be introduced.

Let $\mathcal{D}_t$ be the class of <u>admissible truncated control laws</u>

$$\mathcal{D}_t = \{\{u^{t-1}\} | \{u\} \in \mathcal{D}\} \ .$$

The notation

$$\{u^{t-1}(\cdot) \ , \ u_t(\cdot)\}$$

indicates a truncated law in $\mathcal{D}_{t+1}$, where $\{u^{t-1}\} \in \mathcal{D}_t$ and $u_t(\cdot)$ is a control function. This notation will be used to state a result which follows immediately from (4.2.16) of Therem 4.2.4 and Definition 4.2.4.

Theorem 4.3.5. Let the control system have the countable ϵ-lattice property. Then the conditional loss function $\hat{F}_t$ is determined uniquely by

i) $$\hat{F}_T(z^T,\{u^{T-1}\}) = E_{\{u^{T-1}\}}[L|z^T] \qquad \text{a.s. } P^T_{\{u^{T-1}\}}$$

for $\{u^{T-1}\} \in \mathcal{D}$, and

ii) $\hat{F}_t(z^t,\{u^{t-1}\})$

$$= P^t_{\{u^{t-1}\}}\text{-ess}\inf_{\substack{\bar{u}_t(\cdot) \ni \\ \{u^{t-1}(\cdot),\bar{u}_t(\cdot)\} \in \mathcal{D}_{t+1}}} E_{\{u^{t-1}(\cdot),\bar{u}_t(\cdot)\}}[\hat{F}_{t+1}(z^{t+1},\{u^{t-1}(\cdot),\bar{u}_t(\cdot)\}|z^t] \quad \text{a.s.} P^t_{u^{t-1}}$$

Theorem 4.3.6 i) If the control law $\{\hat{u}\}$ satisfies

(4.3.16) $$\{\hat{u}\} \in \mathcal{D} \ , \ \{\hat{u}^{t-1}\} = \{u^{t-1}\}$$

and

(4.3.17) $$E_{\{\hat{u}^\tau\}}[\hat{F}_{\tau+1}(z^{\tau+1},\{\hat{u}^\tau\})|z^\tau] \le P^\tau_{\{\hat{u}^{\tau-1}\}}\text{- ess}\inf_{\substack{\bar{u}_\tau(\cdot) \ni \\ \{\hat{u}^{\tau-1}(\cdot),u_\tau(\cdot)\} \in \mathcal{D}_{\tau+1}}} E_{\{\hat{u}^{\tau-1}(\cdot),u_\tau(\cdot)\}}[\hat{F}_{\tau+1}(z^{\tau+1},\{\hat{u}^{\tau-1}(\cdot),\bar{u}_\tau(\cdot)\})|z^\tau] + \epsilon_\tau \qquad \text{a.s. } P^\tau_{\{\hat{u}\}}$$

for $\tau = t$, $t+1,\dots,T-1$, then $\{\hat{u}\}$ is ϵ-optimal for $\{u\}$ at t where

(4.3.18) $$\epsilon = \sum_{\tau=t}^{T-1} \epsilon_\tau \ .$$

ii) If the control law $\{\hat{u}\}$ satisfies (4.3.16) and

$$E_{\{\hat{u}^{\tau}\}}[\hat{F}_{\tau+1}(z^{\tau+1},\{\hat{u}^{\tau}\})|z^{\tau}]$$

(4.3.19)$_{\tau}$
$$= P^{\tau}_{\{\hat{u}^{\tau-1}\}}\text{-ess inf}_{\substack{\bar{u}_{\tau}(\cdot)\ \ni \\ \{\hat{u}^{\tau-1}(\cdot),\bar{u}_{\tau}(\cdot)\}\in \mathcal{D}_{\tau+1}}} E_{\{\hat{u}^{\tau-1}(\cdot),\bar{u}_{\tau}(\cdot)\}}[\hat{F}_{\tau+1}(z^{\tau+1},\{\hat{u}^{\tau-1}(\cdot),\bar{u}_{\tau}(\cdot)\})|z^{t}] \quad \text{a.s. } P^{\tau}_{\{\hat{u}\}}$$

for $\tau = t, t+1,\dots,T-1$, the $\{\hat{u}\}$ is optimal for $\{u\}$ at t .

Proof: i) It will be shown by backward induction that

$$\text{(4.3.20)} \qquad E_{\{\hat{u}\}}[L|z^{s}] \leq \hat{F}_{s}(z^{s},\{\hat{u}\}) + \sum_{\tau=s}^{T-1} \epsilon_{\tau}$$

holds for $s = T,T-1,\dots,t$. For $s = T$, from (4.2.6) of Lemma 4.2.1, (4.3.20) holds with equality rather than inequality. Following the usual convention the second term on the right is zero. Assume then that (4.3.20) holds for $s+1$. Taking conditional expectations of the induction hypothesis

$$\text{(4.3.21)} \qquad E_{\{\hat{u}\}}[L|z^{s}] \leq E_{\{\hat{u}\}}[\hat{F}_{s+1}(z^{s+1},\{\hat{u}\})|z^{s}] + \sum_{\tau=s+1}^{T-1} \epsilon_{\tau} \qquad \text{a.s. } P^{s}_{\{\hat{u}\}}$$

From (4.3.21), (4.3.17) for $\tau = s$, and Lemma 4.2.3

$$E_{\{\hat{u}\}}[L|z^{s}] \leq P^{s}_{\{\hat{u}\}}\text{-ess inf}_{\substack{\{\bar{u}\}\in \mathcal{D}\ \ni \\ \{\bar{u}^{t-1}\}=\{\hat{u}^{t-1}\}}} E_{\{\bar{u}\}}[\hat{F}_{s+1}(z^{t+1},\{\bar{u}\})|z^{s}] + \epsilon_{s} + \sum_{\tau=s+1}^{T-1}$$
$$\leq \hat{F}_{s}(z^{t},\{\hat{u}\}) + \sum_{\tau=s}^{T-1} \epsilon_{\tau} \qquad \text{a.s. } P^{t}_{\{\hat{u}\}} \quad .$$

Thus (4.3.20) holds for s . The ϵ-optimality of $\{\hat{u}\}$ follows from (4.3.20) for $s = t$. The proof of ii) is the same as that of i) with $\epsilon_{\tau} = 0$.

Lemma 4.3.1. If the control system has the countable ϵ-lattice property, then for $\{u\} \in \mathcal{D}$ and t fixed the family of functions

$$\text{(4.3.22)} \qquad \{E_{\{\bar{u}\}}[\hat{F}_{t+1}(z^{t+1},\{\bar{u}\})|z^{t}]\}_{\{\bar{u}\}\in \mathcal{D}\ \ni\ \{\bar{u}^{t-1}\}=\{u^{t-1}\}}$$

has the countable ϵ-lattice property in the sense of Definition A.2.3.

Proof: Let $\bar{\mathcal{D}}$ be a countable class of control laws that satisfy

$$(4.3.23) \qquad \{\bar{u}\} \in \mathcal{D} \text{ and } \{\bar{u}^{t-1}\} = \{u^{t-1}\}$$

and let $\varepsilon > 0$. From Theorem 4.3.4 for each $\bar{u} \in \bar{\mathcal{D}}$ there exists $\{\tilde{u}\}$ that is $\frac{\varepsilon}{2}$-optimal for $\{\bar{u}\}$ at $t+1$. Thus $\{\tilde{u}\}$ satisfies

$$(4.3.24) \qquad \{\tilde{u}^t\} = \{\bar{u}^t\} \, , \; \{\tilde{u}\} \in \mathcal{D} \, ,$$

and

$$E_{\{\tilde{u}\}}[L|z^{t+1}] \leq \hat{F}_{t+1}(z^{t+1},\{\bar{u}\}) + \frac{\varepsilon}{2} \qquad \text{a.s. } P^{t+1}_{\{\bar{u}\}} .$$

Taking conditional expectations with respect to $P^t_{\{\bar{u}\}}$

$$(4.3.25) \qquad E_{\{\tilde{u}\}}[L|z^t] \leq E_{\{\bar{u}\}}[\hat{F}_{t+1}(z^{t+1},\{\bar{u}\})|z^t] + \frac{\varepsilon}{2} \qquad \text{a.s. } P^t_{\{\bar{u}\}} .$$

The class $\tilde{\mathcal{D}}$ of laws $\{\tilde{u}\}$ is countable and from (4.3.24) and (4.2.23)

$$\{\tilde{u}^{t-1}\} = \{u^{t-1}\}$$

for all $\{\tilde{u}\} \in \mathcal{D}$. Thus from the countable ε-lattice property (Definition 4.2.6) there exists $\{\tilde{\tilde{u}}\}$ which satisfies

$$(4.3.26) \qquad \{\tilde{\tilde{u}}^{t-1}\} = \{u^{t-1}\} \, , \; \{\tilde{\tilde{u}}\} \in \mathcal{D} \, ,$$

and

$$(4.3.27) \qquad E_{\{\tilde{\tilde{u}}\}}[L|z^t] \leq E_{\{\tilde{u}\}}[L|z^t] + \frac{\varepsilon}{2} \qquad \text{a.s. } P^t_{\{u\}}$$

for all $\{\tilde{u}\} \in \tilde{\mathcal{D}}$. From the definition of $\hat{F}_{t+1}(z^{t+1},\{\tilde{\tilde{u}}\})$

$$\hat{F}_{t+1}(z^{t+1},\{\tilde{\tilde{u}}\})|z^t] \leq E_{\tilde{\tilde{u}}}[L|z^t] \qquad \text{a.s. } P^t_{\{\tilde{\tilde{u}}\}} .$$

Taking conditional expectations

$$(4.3.28) \qquad E_{\tilde{\tilde{u}}}[\hat{F}_{t+1}(z^{t+1},\{\tilde{\tilde{u}}\})|z^t] \leq E_{\tilde{\tilde{u}}}[L|z^t] \qquad \text{a.s. } P^t_{\{u\}} .$$

Thus for $\{\bar{u}\} \in \mathcal{D}$ from (4.3.28), (4.3.27), and (4.3.25)

$$(4.3.29) \qquad \begin{aligned} &E_{\tilde{\tilde{u}}}[\hat{F}_{t+1}(z^{t+1},\{\tilde{\tilde{u}}\})|z^t] \leq E_{\tilde{\tilde{u}}}[L|z^t] \\ &\leq E_{\tilde{u}}[L|z^t] + \frac{\varepsilon}{2} \leq E_{\{\bar{u}\}}[\hat{F}_{t+1}(z^{t+1},\{\bar{u}\})|z^t] + \frac{\varepsilon}{2} + \frac{\varepsilon}{2} \qquad \text{a.s. } P^t_{\{u\}} . \end{aligned}$$

From (4.3.26) the left side of (4.3.29) is a member of the family (4.3.22), and the result follows.

Theorem 4.3.7. Let the system have the countable ϵ-lattice property. Then for all $\{u\} \in \mathcal{D}$, t , and $\epsilon_\tau > 0$, $\tau = t,t+1,\ldots,T-1$, there exists $\{\hat{u}\}$ which satisfies (4.3.16) and (4.3.17).

Proof: The result follows by forward induction from Lemmas 4.3.1 and A.2.16.

While the proof of Theorem 4.3.7 is far from constructive it does assure that if the functionals $\hat{F}_t(z^t,\{u\})$ are known, then ϵ-optimal control laws can be computed step-wise in the forward direction. At each step the P-ess inf is taken with respect to a class of control functions $u_\tau(\cdot)$ rather than control laws.

Relationships between optimality and the martingale properties of $\hat{F}_t$ will be explored next.

Lemma 4.3.2. If $\hat{F}_\tau(z^\tau,\{\hat{u}\})$ is a martingale for $t \leq \tau \leq T$, then $\{\hat{u}\}$ is optimal for $\{u\}$ at τ provided only that $t \leq \tau \leq T$ and $\{u^{\tau-1}\} = \{\hat{u}^{\tau-1}\}$. Further,

$$(4.3.30) \qquad E_{\{\hat{u}\}}[L|z^\tau] = \hat{F}_\tau(z^\tau,\{\hat{u}\}) \qquad \text{a.s. } P^\tau_{\{\hat{u}\}}$$

for $t \leq \tau \leq T$.

Proof: From the martingale property, (4.2.6) of Lemma 4.2.1, and the smoothing property of conditional expectations

$$\hat{F}_\tau(z^\tau,\{\hat{u}\}) = E_{\{\hat{u}\}}[\hat{F}_T(z^T,\{\hat{u}\})|z^\tau]$$

$$= E_{\{\hat{u}\}}[E_{\{\hat{u}\}}[L|z^T]|z^t] = E_{\{\hat{u}\}}[L|z^\tau] \qquad \text{a.s. } P^\tau_{\{\hat{u}\}} .$$

Thus (4.3.30) holds. Optimality of $\{\hat{u}\}$ follows from (4.3.30) and Definition 4.3.1.

The following lemma provides the most elementary and often the most constructive sufficient condition for the optimality of a control law.

<u>Lemma</u> 4.3.3. If there exists a non-negative, compatible functional $F_t(z^t,\{u\})$ that is a clesed, submartingale for all $\{u\} \in \mathcal{D}$ and satisfies

$$(4.3.31) \qquad F_t(z^t,\{\hat{u}\}) = L_{\{\hat{u}\}}[L|z^t] \qquad \text{a.s. } P^t_{\{\hat{u}\}} ,$$

then $\{\hat{u}\}$ is optimal for $\{u\}$ at t provided only that $\{u^{t-1}\} = \{\hat{u}^{t-1}\}$.

Proof: From (4.3.31) , ii) of Theorem 4.2.1, Lemma 4.2.3, Definition 4.2.3, and the definition of $\hat{F}_t(z^t,\{\hat{u}\})$

$$E_{\{\hat{u}\}}[L|z^t] = F_t(z^t,\{\hat{u}\}) \leq \tilde{F}_t(z^t,\{\hat{u}\})$$

$$\leq \hat{F}_t(z^t,\{\hat{u}\}) \leq E_{\hat{u}}[L|z^t] \qquad \text{a.s. } P^t_{\{\hat{u}\}} .$$

Thus

$$E_{\{\hat{u}\}}[L|z^t] = \hat{F}_t(z^t,\{\hat{u}\}) \qquad \text{a.s. } P^t_{\{\hat{u}\}} ,$$

and optimality follows from Definition 4.3.1.

<u>Lemma</u> 4.3.4. If $\hat{F}_t(z^t,\{\hat{u}\})$ is a submartingale and $\{\hat{u}\}$ is optimal for $\{u\}$ at t , then

$$(4.3.32) \qquad \begin{aligned} E_{\{\hat{u}\}}[L|z^t] &= \hat{F}_t(z^t,\{\hat{u}\}) \\ &= E_{\{\hat{u}\}}[\hat{F}_\tau(z^\tau,\{\hat{u}\})|z^t] \qquad \text{a.s. } P^t_{\{\hat{u}\}} \end{aligned}$$

for $\tau = t,t+1,\ldots,T$.

Proof: From the definition of $\hat{F}_\tau(z^\tau,\{\hat{u}\})$

$$\hat{F}_\tau(z^\tau,\{\hat{u}\}) \leq E_{\{\hat{u}\}}[L|z^\tau] \qquad \text{a.s. } P^\tau_{\{\hat{u}\}} .$$

Taking conditional expectations,

$$(4.3.33) \qquad E_{\{\hat{u}\}}[\hat{F}_\tau(z^\tau,\{\hat{u}\})|z^t] \leq E_{\{\hat{u}\}}[L|z^t] \qquad \text{a.s. } P^t_{\{\hat{u}\}} .$$

Thus from Definition 4.3.1, the submartingale property of $\hat{F}_t(z^t,\{\hat{u}\})$ and (4.3.33)

$$E_{\{\hat{u}\}}[L|z^t] = \hat{F}_t(z^t,\{\hat{u}\}) \leq E_{\{\hat{u}\}}[\hat{F}_\tau(z^\tau,\{\hat{u}\})|z^t]$$

$$\leq E_{\{\hat{u}\}}[L|z^t] \qquad \text{a.s. } P^t_{\{\hat{u}\}}$$

and (4.3.32) follows.

Lemma 4.3.5. Let $\hat{F}_\tau(z^\tau,\{\hat{u}\})$ be a submartingale where $\{\hat{u}\}$ is optimal for some $\{u\}$ at t and

(4.3.34) $$E_{\{\hat{u}\}}[L] < \infty \ .$$

Then $\hat{F}_\tau(z^\tau),\{\hat{u}\})$ is a martingale for $t \leq \tau \leq T$, and $\{\hat{u}\}$ is optimal for $\{u\}$ at τ provided only that $\{u^{\tau-1}\} = \{\hat{u}^{\tau-1}\}$ and $t \leq \tau \leq T$.

Proof: From the previous lemma for $t \leq \tau \leq T$

(4.3.35) $$E_{\{\hat{u}\}}[E_{\{\hat{u}\}}[L|z^\tau]|z^t] \leq E_{\{\hat{u}\}}[\hat{F}_\tau(z^\tau,\{\hat{u}\})|z^t] \qquad \text{a.s. } P^t_{\hat{u}} \ .$$

From the definition of $\hat{F}_\tau(z^\tau,\{\hat{u}\})$ and (4.3.34)

(4.3.36) $$\hat{F}_\tau(z^\tau,\{\hat{u}\}) \leq E_{\{\hat{u}\}}[L|z^\tau] < \infty \qquad \text{a.s. } P^\tau_{\{\hat{u}\}} \ .$$

From (4.3.35) and the finiteness of (4.3.36)

(4.3.37) $$0 \leq E_{\{\hat{u}\}}\{\hat{F}_\tau(z^\tau,\{\hat{u}\}) - E_{\{\hat{u}\}}[L|z^\tau]\} \ .$$

Since the random variable { } is non-positive, it follows from (4.3.37) that it must be a.s. zero. Thus

$$\hat{F}_\tau(z^\tau,\{\hat{u}\}) = E_{\{\hat{u}\}}[L|z^\tau] \qquad \text{a.s. } P^\tau_{\{\hat{u}\}} \ .$$

The martingale property of $\hat{F}_\tau(z^\tau,\{\hat{u}\})$ is inherited from that of $E_{\{\hat{u}\}}[L|z^\tau]$.

Theorem 4.3.8. Let the control system have the countable ϵ-lattice property. Then $\{\hat{u}\} \in \mathcal{D}$ is optimal for $\{u\}$ at t for all t and $\{u\}$ satisfying

$$(4.3.38) \qquad \{u^{t-1}\} = \{\hat{u}^{t-1}\}$$

if and only if there exists a non-negative, compatible functional $F_t(z^t,\{u\})$ which is a closed submartingale for all $\{u\} \in \mathcal{D}$ and a martingale for $\{\hat{u}\}$.

Proof: If such a functional $F_t(z^t,\{u\})$ exists, then since $F_t(z^t\{\hat{u}\})$ is a closed martingale

$$F_t(z^t,\{\hat{u}\}) = E_{\{\hat{u}\}}[L|z^t] \qquad \text{a.s. } P^t_{\{\hat{u}\}} ,$$

and optimality of $\{\hat{u}\}$ follows from Lemma 4.3.3, From Theorem 4.2.5 i) $F_t = \hat{F}_t$ is a non-negative, compatible functional which is a closed submartingale for all $\{u\}$. If $\{\hat{u}\}$ has the optimality property of the theorem, then from Lemma 4.3.4

$$E_{\{\hat{u}\}}[L|z^t] = \hat{F}_t(z^t,\{\hat{u}\}) \qquad \text{a.s. } P^t_{\{\hat{u}\}}$$

for all t. Thus the functional $\hat{F}_t(z^t,\{\hat{u}\})$ is a martingale since the conditional expectation is a martingale.

4.4 Minimum loss function

In this section the possibility of replacing the $P^t_{\{u\}}$-ess inf in the definition of $\hat{F}_t(z^t,\{u\})$ by the usual infimum will be explored; optimality (in the sense of Strauch [16]) with respect to the minimum loss function will be compared with the optimality of Definitions 4.3.1 and 4.3.2 ; and conditions will be given under which the conditional loss functional depends on the law $\{u^{t-1}\}$ only thru its value $u^{t-1}(z^{t-1})$ at the observation point z^t.

First, it should be noticed that it hardly makes sense to replace the P-ess inf in (4.2.1) of Definition 4.2.3 by the usual infimum for each

z^t fixed since the conditional expectations

$$E_{\{\bar{u}\}}[L|z^t]$$

are unique only up to sets of measure zero. This ambiguity is overcome if conditional expectations are computed by integrating with respect to a regular conditional distributions. (See, for example, Hinderer [8] Appendix 3.) Since the incomplete information problem is considered here, an extension of the filtering results of section 2.2 will be required in addition to the standard Markov property of the model.

Lemma 4.4.1. Let $\mathcal{X}^t$ and $\mathcal{Z}^t$ be finite dimensional Euclidean spaces. Then there exist stochastic kernels $\bar{G}_t(z^t,u^{t-1};A)$ on $(\mathcal{Z}^t \times \mathcal{U}^{t-1},\mathcal{X}^t)$ that satisfies

$$(4.4.1) \qquad P_{\{u\}}[x^t \in A|z^t] = \bar{G}_t(z^t,u^{t-1}(z^{t-1});A) \qquad \text{a.s. } P^t_{\{u\}}$$

for all control laws $\{u\}$ and $A \in \mathcal{B}^t_{\mathcal{X}}$.

Proof: This can be obtained from results of section 2.2 for a slightly altered model. Replace x_t by x^t , let

$$x^{t+1} = \bar{\varphi}_t(x^t,u_t,w_{t+1})$$

where the components $\bar{\varphi}^s_t$ of $\bar{\varphi}_t$ are given by

$$\bar{\varphi}^s_t(x^t,u_t,w_{t+1}) = x_s \qquad s = 0,1,\ldots,t$$

$$\bar{\varphi}^{t+1}_t(x^t,u_t,w_{t+1}) = \varphi_t(x_t,u_t,w_{t+1}) ,$$

and let

$$z_t = \bar{\psi}_t(x^t,u_{t-1},e_t) = \psi_t(x_t,u_{t-1},e_t) .$$

This model has the same form as (1.2.1)-(1.2.4), and the result follows from Theorem 2.2.3.

Lemma 4.4.2. For control functions $u_t(\cdot),\ldots,u_{T-1}(\cdot)$ there exists $\bar{\mathcal{L}}_t(x^t,z^t,u^{t-1};u_t(\cdot),\ldots,u_{T-1}(\cdot))$ measurable in (x^t,z^t,u^{t-1}) that satisfies

$$E_{\{u\}}[L|x^t,z^t] = \bar{\mathcal{L}}_t(x^t,z^t,u^{t-1}(z^{t-1});u_t(\cdot),\ldots,u_{T-1}(\cdot)) \quad \text{a.s. } P_{\{u\}} \tag{4.4.2}$$

for all control laws $\{u\}$.

Proof: For t and $u_t(\cdot),u_{t+1}(\cdot),\ldots,u_{T-1}(\cdot)$ given define functions $\bar{U}_{\tau-1},\bar{X}_\tau$, and $\bar{z}_\tau$ for $\tau > t$ by successive substitution from the model

$$\begin{aligned}
&\bar{U}_t(z^t) \doteq u_t(z^t)\\
&\bar{X}_{t+1}(x_t,z^t,w_{t+1}) = \varphi_t(x_t,\bar{U}_t(z^t),w_{t+1})\\
&\bar{Z}_{t+1}(x_t,z^t,w_{t+1},e_{t+1}) = \psi_{t+1}(\bar{X}_{t+1}(x_t,z^t,w_{t+1}),\bar{U}_t(z^t),e_{t+1})\\
&\vdots\\
&\bar{U}_{\tau-1}(x^t,z^t,e_{t+1},\ldots,e_{\tau-1},w_{t+1},\ldots,w_{\tau-1})\\
&\qquad = u_{\tau-1}(z^t,\bar{Z}_{t+1},\bar{Z}_{t+2},\ldots,\bar{Z}_{\tau-1})\\
&\bar{X}_\tau(x_t,z^t,e_{t+1},\ldots,e_{\tau-1},w_{t+1},\ldots,w_\tau)\\
&\qquad = \varphi_{\tau-1}(\bar{X}_{\tau-1},\bar{U}_{\tau-1},w_\tau)\\
&\bar{Z}_\tau(x_t,z^t,e_{t+1},\ldots,e_\tau,w_{t+1},\ldots,w_\tau)\\
&\qquad = \psi_\tau(\bar{X}_\tau,\bar{U}_{\tau-1},e_\tau)\\
&\vdots
\end{aligned} \tag{4.4.3}$$

It is clear from the construction that the random variables defined by (1.2.13) then satisfy

$$\begin{aligned}
u_{\tau-1} &= \bar{U}_{\tau-1}(X_t^{\{u^{t-1}\}}(e^{t-1},w^t),Z^{t,\{u^{t-1}\}}(e^t,w^t)\ ,\\
&\qquad e_{t+1},\ldots,e_{\tau-1},w_{t+1},\ldots,w_{\tau-1})\\
x_\tau &= \bar{X}_\tau(X_t^{\{u^{t-1}\}}(e^{t-1},w^t),Z^{t,\{u^{t-1}\}}(e^t,w^t),e_{t+1},\ldots,e_{\tau-1},w_{t+1},\ldots,w_\tau)\\
z_\tau &= Z_\tau(X_t^{\{u^{t-1}\}}(e^{t-1},w^t),Z^{t,\{u^{t-1}\}}(e^t,w^t),e_{t+1},\ldots,e_{\tau-1},w_{t+1},\ldots,w_\tau)\ .
\end{aligned}$$

Lemma A.1.8 will be applied with

$$\begin{aligned}
X &= (x^t,z^t,u^{t-1})\\
Y &= (x^T,z^T,u^{T-1})\\
w' &= (e^t,w^t)\\
w'' &= (e_{t+1},\ldots,e_T,w_{t+1},\ldots,w_T)\ ,
\end{aligned}$$

$\psi(x,w'')$ defined by (4.4.3) for $\tau = t+1,\ldots,T$, $X(w')$ defined by (1.2.13) for $\tau = 1,2,\ldots, t$. Then from the lemma

$$(4.4.4)\quad \begin{aligned} H(x^t,z^t,u^{t-1};B) = P_\Omega[\{w''|x,\bar{X}_{t+1}(x,w''),\ldots,\bar{X}_\tau(x,w''),\\ \bar{Z}_t(x,w''),\ldots,\bar{Z}_\tau(x,w''),\bar{U}_t(x,w''),\ldots,\bar{U}_{T-1}(x,w''))\in B] \end{aligned}$$

is a stochastic kernel on $(\mathcal{X}^t \times Z^t \times \mathcal{U}^{t-1}, \mathcal{X}^T \times Z^T \times \mathcal{U}^{T-1})$ and satisfies

$$H(x^t,z^t,u^{t-1};B) = P_{\{u\}}[(x^T,z^T,u^{T-1}) \in B|x^t,z^t,u^{t-1}] \qquad \text{a.s. } P_{\{u\}} .$$

Further, from (4.4.4) H depends only on the law only thru the function (4.4.3) and hence only thru the functions $u_t(\cdot),\ldots,u_{T-1}(\cdot)$. Let

$$\bar{\mathcal{L}}_t(x^t,z^t,u^{t-1};u_t(\cdot),\ldots,u_{T-1}(\cdot)) = \int_{\mathcal{X}^T \times Z^T \times \mathcal{U}^{T-1}} L(x^T,z^T,u^{T-1})H(x^t,z^t,u^{t-1};dx^T,d\tau^T,du^{T-1}).$$

The measurability of $\bar{\mathcal{L}}_t$ follows from Lemma A.1.1, and (4.4.2) follows from Lemma A.1.5.

<u>Lemma</u> 4.4.3. For all control functions $u_t(\cdot),\ldots,u_{T-1}(\cdot)$ there exists $\mathcal{L}_t(z^t,u^{t-1};u_t(\cdot),\ldots,u_{T-1}(\cdot))$ measurable in (z^t,u^{t-1}) which satisfies

$$(4.4.5)\qquad E_{\{u\}}[L|z^t] = \mathcal{L}_t(z^t,u^{t-1}(z^{t-1});u_t(\cdot),\ldots,u_{T-1}(\cdot)) \qquad \text{a.s. } P^t_{\{u\}}$$

for all control laws $\{u\}$.

Proof: Letting

$$\mathcal{L}_t(z^t,u^{t-1};u_t(\cdot),\ldots,u_{T-1}(\cdot)) = \int_{\mathcal{X}^t} \bar{\mathcal{L}}_t(x^t,z^t,u^{t-1};u_t(\cdot),\ldots,u_{T-1}(\cdot))\bar{G}_t(z^t,u^{t-1};dx^t),$$

measurability of $\mathcal{L}_t$ follows from Lemma A.1.1. From (4.4.2), (4.4.1), and Lemma A.1.5

$$\begin{aligned} E_{\{u\}}[L|z^t] &= E_{\{u\}}[E_{\{u\}}[L|x^t,z^t,u^{t-1}]|z^t] \\ &= E_{\{u\}}[\bar{\mathcal{L}}_t(x^t,z^t,u^{t-1}(z^{t-1});u_t(\cdot),\ldots,u_{T-1}(\cdot))|z^t] \\ &= \mathcal{L}_t(z^t,u^{t-1}(z^{t-1});u_t(\cdot),\ldots,u_{T-1}(\cdot)) \qquad \text{a.s. } P^t_{\{u\}} . \end{aligned}$$

Definition 4.4.1. The minimum loss function is defined by

$$(4.4.6) \qquad \hat{\mathcal{F}}_t(z^t,u^{t-1};\{u\}) = \inf_{\substack{\{\bar{u}\}\in\mathcal{D} \ni \\ \{\bar{u}^{t-1}\}=\{u^{t-1}\}}} \mathcal{L}_t(z^t,u^{t-1};\bar{u}_t(\cdot),\ldots,\bar{u}_{T-1}(\cdot))$$

for $(z^t,u^{t-1}) \in Z^t \times U^{t-1}$ and $\{u\} \in \mathcal{D}$.

It should be noted that $\hat{\mathcal{F}}_t$ depends only on $\{u^{t-1}\}$, but that it is not necessarily measurable in (z^t,u^{t-1}) .

Lemma 4.4.4. If there exists $\hat{\mathcal{F}}^*_t(z^t,u^{t-1};\{u\})$ measurable in (z^t,u^{t-1}) which depends on $\{u\}$ only thru $\{u^{t-1}\}$ and satisfies

$$(4.4.7) \qquad \hat{\mathcal{F}}^*_t(z^t,u^t(z^{t-1});\{u\}) = \hat{\mathcal{F}}_t(z^t,u^{t-1}(z^{t-1});\{u\}) \qquad \text{a.s. } P^t_{\{u\}}$$

for all $\{u\} \in \mathcal{D}$, then

$$(4.4.8) \qquad \hat{\mathcal{F}}^*_t(z^t,u^{t-1}(z^{t-1});\{u\}) \le \hat{F}_t(z^t,\{u\}) \qquad \text{a.s. } P^t_{\{u\}}.$$

Proof: This follows from Lemmas A.2.18 and 4.4.3 and Definitions 4.2.3 and 4.4.1.

In general no such measurable $\hat{\mathcal{F}}^*_t$ need exist, and if it does exist, the inequality (4.4.8) may be strict. The example following Lemma A.2.18 can be translated into a control system by letting

$$T = 2$$

$$Z_1 = \epsilon_1 = \omega$$

$$L = u_1$$

and letting the class of admissible control laws $\mathcal{D}$ be the family of function $\{f_\gamma(\omega)\}$ of the example.

Definition 4.4.2. The control law $\{\hat{u}^\epsilon\}$ is ϵ-optimal in the sense of Strauch for $\{u\}$ at t provided $\{\hat{u}^\epsilon\} \in \mathcal{D}$, $\{\hat{u}^{t-1,\epsilon}\} = \{u^{t-1}\}$, and

$$(4.4.9) \quad \mathcal{L}_t(z^t,u^{t-1}(z^{t-1});\hat{u}^\epsilon_t(\cdot),\ldots,\hat{u}^\epsilon_{T-1}(\cdot)) \le \hat{\mathcal{F}}_t(z^t,u^{t-1}(z^{t-1});\{u\}) + \epsilon \quad \text{a.s. } P^t_{\{u\}} \ .$$

The exceptional set in (4.4.9) will not in general be measurable. The requirement (4.4.9) is that it be contained in a measurable set of $P^t_{\{u\}}$-measure zero.

Lemma 4.4.5. If $\{\hat{u}^\epsilon\}$ is ϵ-optimal in the sense of Stauch for $\{u\}$ at t, then $\{\hat{u}^\epsilon\}$ is ϵ-optimal for $\{u\}$ at t.

Proof: From Lemma 4.4.3, (4.4.9), and Definition 4.4.1

$$E_{\{\hat{u}^\epsilon\}}[L|z^t] = \mathcal{L}_t(z^t,u^{t-1}(z^{t-1});\hat{u}^\epsilon_t(\cdot),\ldots,\hat{u}^\epsilon_{T-1}(\cdot))$$

$$\leq \hat{\mathcal{F}}_t(z^t,u^{t-1}(z^{t-1});\{u^{t-1}\}) + \epsilon \leq \mathcal{L}_t(z^t,u^{t-1}(z^{t-1});\bar{u}_t(\cdot),\ldots,\bar{u}_{T-1}(\cdot)) + \epsilon$$

$$= E_{\{\bar{u}\}}[L|z^t] + \epsilon \qquad \text{a.s. } P^t_{\{u\}}$$

for all $\{\bar{u}\} \in \mathcal{D}$ such that $\{\bar{u}^{t-1}\} = \{u^{t-1}\}$. It follows from Definitions A.1.1 and 4.2.3 that

$$E_{\{\hat{u}^\epsilon\}}[L|z^t] \leq P^t_{\{u\}}\text{-ess inf}_{\substack{\{\bar{u}\} \in \mathcal{D} \ni \\ \{\bar{u}^{t-1}\} = \{u^{t-1}\}}} E_{\{\bar{u}\}}[L|z^t] + \epsilon$$

$$= \hat{F}_t(z^t,\{u\}) + \epsilon \qquad \text{a.s. } P^t_{\{u\}}$$

and hence from Definition 4.3.2 that $\{\hat{u}^\epsilon\}$ is ϵ-optimal for $\{u\}$ at t.

Theorem 4.4.1. If for all $\epsilon > 0$ there exists $\{\hat{u}^\epsilon\}$ ϵ-optimal in the sense of Strauch for $\{u\}$ at t, then

$$\hat{F}_t(z^t,\{u\}) = \hat{\mathcal{F}}_t(z^t,u^{t-1}(z^{t-1});\{u^{t-1}\}) \qquad \text{a.s. } P^t_{\{u\}} \tag{4.4.10}$$

and a control law is ϵ-optimal for $\{u\}$ at t if and only if it is ϵ-optimal in the sense of Strauch.

Proof: The result (4.4.10) follows from Lemmas A.2.19 and 4.4.3 and the definitions of $\hat{F}_t$ and $\hat{\mathcal{F}}_t$ (4.2.3 and 4.4.1). The equivalence of ϵ-optimal properties follows from (4.4.10), Lemma 4.4.3 and the definitions (4.3.2 and 4.4.2) of ϵ-optimality.

Is is clear that optimal (Definition 4.3.1) and optimal in the sense of Struach are also equivalent under the assumptions of Theorem 4.4.1.

If the class of admissible laws $\mathfrak{D}$ contains all measurable laws then the infimum in (4.4.6) is taken over all control functions $\{\bar{u}_t(\cdot),\ldots,\bar{u}_{T-1}(\cdot)\}$, and $\hat{\mathfrak{F}}_t$ is independent of $\{u^{t-1}\}$. Thus $\hat{\mathfrak{F}}_t(z^t,u^{t-1})$ is a function instead of a functional. If $\hat{\mathfrak{F}}_t(z^t,u^{t-1})$ is universally measurable and there exist ϵ-optimal controls in the sense of Strauch (see Strauch [16] Th. 8.1 and Hinderer[8] Th. 14.1), then from Theorem 4.4.1

$$\hat{F}_t(z^t,\{u\}) = \hat{\mathfrak{F}}_t(z^t,u^{t-1}(z^{t-1})) \qquad \text{a.s. } P^t_{\{u\}} \quad .$$

If the class of terminal controls $\{u_t(\cdot),\ldots,u_{T-1}(\cdot)\}$ for which $\{u^{t-1}(\cdot),u_t(\cdot),\ldots,u_{T-1}(\cdot)\} \in \mathfrak{D}$ depends on $\{u^{t-1}\}$, then dependence of $\hat{\mathfrak{F}}_t$ on $\{u^{t-1}\}$ cannot be avoided. Defining $\hat{\mathfrak{F}}_t$ by

$$\hat{\mathfrak{F}}_t(z^t,u^{t-1}) = \inf_{\substack{\{u\} \in \mathfrak{D} \ni \\ u^{t-1}(z^{t-1}) = u^{t-1}}} \mathcal{L}_t(z^t,u^{t-1};u_t(\cdot),\ldots,u_T(\cdot))$$

in this case hardly seems reasonable, since the infimum is taken in a non-trivial way over a class of truncated laws $\{u^{t-1}\}$. At time t, $\{u^{t-1}\}$ is over with; even if the law $\{u^{t-1}\}$ that was used is unknown, a loss criterion that involves minimization over the past is hardly acceptable.

Blackwell[2], Strauch [16], and Hinderer [8] make use of properties of the class of all control laws that is very closely related to the ϵ-lattice property. Hinderer [8] calls these properties stable and σ-stable.

<u>Definition</u> 4.4.3. Let $\{\bar{u}\}$ and $\{\bar{\bar{u}}\}$ be control laws which satisfy

$$\{\bar{u}^{t-1}\} = \{\bar{\bar{u}}^{t-1}\} \tag{4.4.11}$$

Then $\{\tilde{u}\}$ is <u>generated</u> by $\{\bar{u}\}$ and $\{\bar{\bar{u}}\}$ provided from some $A^t \in \mathcal{B}^t_Z$

$$\{\tilde{u}^{t-1}\} = \{\bar{u}^{t-1}\} \tag{4.4.12}$$

$$\tilde{u}_\tau(z^\tau) = \chi_{A^t \times Z_{t+1} \times \ldots \times Z_\tau}(z^\tau)\, \bar{u}_\tau(z^\tau) + \chi_{(A^t)^c \times Z_{t+1} \times \ldots \times Z_\tau}(z^\tau)\, \bar{\bar{u}}_\tau(z^\tau) \tag{4.4.13}$$

$\tau = t, t+1, \ldots,$. The class $\mathcal{D}$ is stable provided $\{\tilde{u}\}$ generated by $\{\bar{u}\} \in \mathcal{D}$ and $\{\bar{\bar{u}}\} \in \mathcal{D}$ implies that $\{\tilde{u}\} \in \mathcal{D}$.

Definition 4.4.4. Let $\{u^n\}$, $n = 1,2,\ldots$ be control laws which satisfy

$$\{u^{t-1,n}\} = \{u^{t-1,m}\} \tag{4.4.14}$$

for all n,m . Then $\{\tilde{u}\}$ is generated by the countable class $\{u^n\}$ provided

$$\{\tilde{u}^{t-1}\} = \{u^{t-1,n}\}$$

and there exist sets $A_n^t \in \mathcal{B}_Z^t$ which satisfy

$$\sum_{n=1}^{\infty} A_n^t = Z^t \tag{4.4.15}$$

and

$$\tilde{u}_\tau(z^\tau) = \sum_{n=1}^{\infty} \chi_{A_n^t \times Z_{t+1} \times \ldots \times Z_\tau}(z^\tau) u_\tau^n(z^\tau) \tag{4.4.16}$$

for $\tau = t, t+1, \ldots$. The class $\mathcal{D}$ is σ-stable provided $\{\tilde{u}\}$ generated by $\{u^n\}$ where $\{u^n\} \in \mathcal{D}$ implies that $\{\tilde{u}\} \in \mathcal{D}$.

Lemma 4.4.6. Let $A^t \in \mathcal{B}_Z^t$, $\{u\}$ and $\{\bar{u}\}$ satisfy

$$\{u^{t-1}\} = \{\bar{u}^{t-1}\} \tag{4.4.17}$$

$$u_\tau(z^\tau) = \bar{u}_\tau(z^\tau) \quad \text{for } z^\tau \in A^t \times Z_{t+1} \times \ldots \times Z_\tau \tag{4.4.18}$$

for $\tau = t, t+1, \ldots$ and let $L(x^T, z^T, u^{T-1}) \geq 0$ be measurable. Then

$$\chi_{A^t}(z^t) E_{\{u\}}[L|z^t] = \chi_{A^t}(z^t) E_{\{\bar{u}\}}[L|z^t] \quad \text{a.s. } P_{\{u\}}^t . \tag{4.4.19}$$

Proof: From the substitution (1.2.13) and the conditions (4.4.17) and (4.4.18) it can easily be seen that for $z^t \in A^t$, the functions $U_{\tau-1}^{\{u\}}(\omega)$, $X_\tau^{\{u\}}(\omega)$, $Z_\tau^{\{u\}}(\omega)$ and $U_{\tau-1}^{\{\bar{u}\}}(\omega)$, $X_\tau^{\{\bar{u}\}}(\omega)$, $Z_\tau^{\{\bar{u}\}}(\omega)$ are identical in ω for $\tau = 1,2,\ldots,T$. Thus

$$\chi_{A^t}(Z^{t,\{u\}}(\omega))L(X^{T,\{u\}}(\omega),\ Z^{T,\{u\}}(\omega),U^{T-1,\{u\}}(\omega))$$

$$= \chi_{A^t}(Z_t^{\{\bar{u}\}}(\omega))L(X^{T,\{\bar{u}\}}(\omega),Z^{T,\{u\}}(\omega),U^{T-1,\{\bar{u}\}}(\omega))$$

for all $\omega \in \Omega$, and (4.4.19) follows from the definition of the distributions $P_{\{u\}}$ and $P_{\{\bar{u}\}}$.

Theorem 4.4.2. i) If the class $\mathfrak{D}$ is stable, then the control system has the finite ε-lattice property.

ii) If the class $\mathfrak{D}$ is σ-stable, then the system has the countable ε-lattice property.

Proof: Only ii) will be proved. Let $\bar{\mathfrak{D}}$ be a countable subclass which satisfies (4.2.10) of Definition 4.2.6. Order the class $\bar{\mathfrak{D}} = \{\{u^n\}\} n=1,2,\dots$ and for $\varepsilon > 0$ define A_n^t inductively $n = 1,2,\dots$ by

$$A_n^t = \{z^t | z^t \notin \bigcup_{k=1}^{n-1} A_k ,\ E_{\{u^n\}}[L|z^t] \leq \inf_{\{\bar{u}\}\in\mathfrak{D}} E_{\{\bar{u}\}}[L|z^t] + \varepsilon\} . \tag{4.4.20}$$

From the definition (4.4.20)

$$A_n^t \in \mathcal{B}_Z^t \quad , \quad \sum_{n=1}^{\infty} A_n^t = Z^t \quad .$$

Let $\{\tilde{u}\}$ be defined by (4.4.16). Then from Lemma 4.4.6 and (4.4.20) for $z^t \in A_n^t$

$$E_{\{\tilde{u}\}}[L|z^t] = E_{\{u^n\}}[L|z^t]$$

$$\leq \inf_{\{\bar{u}\}\in\bar{\mathfrak{D}}} E_{\{\bar{u}\}}[L|z^t] + \varepsilon \qquad \text{a.s. } P_{\{u\}}^t$$

and (4.2.12) of Definition 4.2.6 follows.

Chapter 5 - Selection Classes

5.1 Introduction

In this chapter solution of the dynamic programming equation will be exploited as a constructive method for obtaining optimal control laws. The principal thrust is the development of optimality properties of laws obtained by the dynamic programming method. While some existence results will be given, they should be considered as peripheral to the main purpose. For example, the results of section 5.4 (principally Theorems 5.4.2 and 5.4.3) can be loosely summarized as stating that, while truly satisfactory laws may or may not exist, the laws obtained by the dynamic programming method are the best possible.

Throughout this chapter it is assumed that Y is a statistic sufficient for V (Definitions 3.2.2 and 3.2.6) and that the loss function has structure V - that is, L satisfies (3.2.7). Under assumptions which guarantee the existence of stochastic kernels G_t and K_t which satisfy (2.1.1) and (2.1.2), it follows from Lemma 3.2.4 that there always exists a sufficient statistic for loss functions of the form (1.2.32). Thus under these assumptions (see Theorems 2.2.3 and 2.3.1) restricting attention to systems with a sufficient statistic implies no loss of generality.

From the assumption of a sufficient statistic, stochastic kernels $G_t^{Y,V}$ and K_t^Y satisfying (3.2.8) and (3.2.2) are available. Making use of these kernels and the loss function satisfying (3.2.7), the constructive solution of the dynamic programming equation which will be used in this chapter can be described heuristically as follows. First, the cost functions c_t^V of (3.2.7) are replaced by the conditional cost functions c_t^Y computed by (5.2.2). Starting at $t = T-1$, control functions $\hat{U}_t(y_t)$ are computed by a backward interation. Compute

$$(5.1.1)\quad L_{T-1}(y_{T-1},u_{T-1}) = \int c_T^Y(y_T,u_{T-1})K_{T-1}^Y(y_{T-1},u_{T-1};dy_T) .$$

For y_{T-1} fixed, the function $L_{T-1}(y_{T-1},u_{T-1})$ is to be minimized in u_{T-1} by the function $\hat{U}_{T-1}(y_{T-1})$ in the following sense. If u^* can be found that minimizes L_{T-1}-that is,

(5.1.2) $$L_{T-1}(y_{T-1},u^*) = \inf_{u_{T-1} \in U_{T-1}} L_{T-1}(y_{T-1},u_{T-1}) ,$$

then let

(5.1.3) $$\hat{U}_{T-1}(y_{T-1}) = u^* .$$

If no such u^* exists, then a compromise is made by letting

(5.1.4) $$\hat{U}_{T-1}(y_{T-1}) = u^{**}$$

where u^{**} satisfies

(5.1.5) $$L_{T-1}(y_{T-1},u^{**}) \leq \inf_{u_{T-1} \in \mathcal{U}_{T-1}} L_{T-1}(y_{T-1},u_{T-1}) + \varepsilon$$

for some $\varepsilon > 0$.

The difficulty with this procedure is that the resulting function $\hat{U}_{T-1}(y_{T-1})$ may not be measurable in y_{T-1} . By application of the appropriate selection theorem it is known that $\hat{U}_{T-1}(y_{T-1})$ can be "selected" is such a way that it is universally measurable provided $L_{T-1}(y_{T-1},u_{T-1})$ is jointly Borel measurable. (This point is discussed in the proof of Lemma 5.4.1.) While such a universally measurable function $\hat{U}_{T-1}$ is quite acceptable as a control function, difficulties arise when the procedure is iterated. If $\hat{U}_{T-1}(y_{T-1})$ is universally measurable ,

(5.1.6) $$\ell_{T-1}(y_{T-1}) = L_{T-1}(y_{T-1},\hat{U}_{T-1}(y_{T-1})) ,$$

and

(5.1.7) $$L_{T-2}(y_{T-2},u_{T-2}) = \int[\ell_{T-1}(y_{T-1}) + c^Y_{T-1}(y_{T-1},u_{T-2})]K^Y_{T-2}(y_{T-2},u_{T-2};dy_{T-1}) ,$$

the function L_{T-2} is universally measurable rather that Borel measurable. Thus the selection theorem no longer applies. This difficulty is dealt with by the introduction of selection classes $\mathcal{D}_0$ of control laws. Instead of settling for $\hat{U}_{T-1}(y_{T-1})$ universally measurable, it is required to be Borel measurable. In order to accomplish this, the requirements (5.1.2)-(5.1.5) are relaxed. These equations are not required to hold for all y_{T-1} but rather a.s. $(Y^{-1}P)^{T-1}_{\{u\}}$ for $\{u\} \in \mathcal{D}_0$. The intention, of course, is that the selection class $\mathcal{D}_0$ be as large as possible. This idea is made precise in Definition 5.2.3. In Theorem 5.4.8, which may be taken as the motivation for Definition 5.2.3, it is shown that $\mathcal{D}_\rho$ (Definition 3.2.10) is a selection class.

The important case, of course, is that of the single selection class, $\mathcal{D}_0 = \mathcal{D}$. In Theorem 5.5.1, it is shown that this holds provided the statistic Y is dominated (Definition 3.2.11) and the spaces $\mathcal{Y}_t$ and $\mathcal{U}_t$ are complete separable metric spaces. Theorems 5.5.6 and 5.5.7 treat the cases in which the spaces $\mathcal{Y}_t$ and $\mathcal{U}_t$ are countable. Another approach to showing the existence of a single selection class is through the imposition of continiuty assumptions. This method, used by Maitra [13] and Hinderer [8] (section 17) is beyond the scope of the present work.

The general method then consists in contructing the laws $\{\hat{U}^{Y,\mathcal{D}_0,\underline{\varepsilon}}\}$ according to the prescriptions of Definition 5.2.3 for as wide a class $\mathcal{D}_0$ as possible with the hope that $\mathcal{D}_0 = \mathcal{D}$. As a practical matter, assumption of additional properties such as the continuity conditions of Maitra [13] or the absolute continuity condition of Theorem 5.5.1 which imply a single selection class is not necessary since in applications in which there is any hope of actually constructing an optimal control law, the question of the existence of a single selection class will be answered easily by direct computation from the model.

The existence results of Theorems 5.4.8 and 5.5.1 are proved under the assumption that the class of admissible laws $\mathcal{D}$ is universal in Y (Definition 5.4.3). While this is a weaker assumption than stability (Definition 4.4.4), the difference is not really important. Universal in Y is essentially the appropriate formulation of the stability property in the presence of a sufficient statistic.

In section 4.3 it was seen that ε-optimal controls always exist provided the system has the countable ε-lattice property. However, the law $\{\hat{u}^\varepsilon\}$ that is ε-optimal for $\{u\}$ at t will in general depend on the truncated law $\{u^{t-1}\}$. With the use of sufficient statistics, a stronger type of optimality is possible. Control laws in Y introduce dependence on the previous controls $\{u^{t-1}\}$ in a natural way since $U_t(Y_t(z^t,u^{t-1}(z^{t-1}))$ depends on $\{u^{t-1}\}$ through

Y_t . This is an important property of sufficient statistics in addition to the usual data reduction property. A control law $\{\hat{U}\}$ is said to be a universally optimal law provided what ever law $\{u\}$ is used up to time t changing to $\{\hat{U}\}$ gives a law optimal for $\{u\}$ at t . This is stated precisely in Definitions 5.5.2 and 5.5.3 .

The definition and properties of a single strong selection class are given at the end of section 5.5. There are the results that are used in the examples of Chapters 6 and 7. No existence results are given for the strong selection class since that would necessitate the introduction of topological assumptions which is outside the range of this work. In the examples considered the existence of the single strong selection class follows easily by direct computation. It is the optimality properties of such a system, stated in Theorem 5.5.5, that are of use in the applications.

Results of Strauch [16] in the stationary case and Hinderer [8] in the non-stationary case are related to some of the results of this chapter. They are concerned primarily with functions $\hat{L}_t(y_t,u_t)$ and $\hat{\ell}_t(y_t)$ which satisfy (5.2.3) and (5.2.32) of Lemma 5.2.4. In Theorems 13.2 and 14.4 Hinderer [8] shows that there exists $\hat{L}_t$ and $\hat{\ell}_t$ universally measurable which satisfy (5.2.3) and (5.2.32) for all y_t . It can easily be shown that the functions $\hat{\ell}_t^{Y,\mathcal{D}_0}$ of Lemma 5.2.4 satisfy

$$(5.1.8) \qquad \hat{\ell}_t^{Y,\mathcal{D}_0}(y_t) = \hat{\ell}_t(y_t) \qquad \text{a.s. } (Y^{-1}P)^t_{\{u\}}$$

for $\{u\} \in \mathcal{D}_0$. Theorem 17.6 of Hinderer [8] states that $\{\hat{u}\}$ is optimal if and only if it satisfies

$$(5.1.9) \qquad \hat{\ell}_t(y_t) = \hat{L}_t(y_t,\hat{u}_t(z^t)) \qquad \text{a.s. } (Y^{-1}P)^t_{\{\hat{u}\}} \quad .$$

This is essentially the same as Theorem 5.4.6. These two results provide a constructive method for obtaining $\{\hat{u}\}$ (provided such a law exists) very similar to method of Theorem 4.3.7. (See the discussion following Theorem 4.3.7) Once the functions $\hat{L}_t$ and $\hat{\ell}_t$ have been obtained by backward induction from (5.2.3) and (5.2.32), then the law $\{\hat{u}\}$ can be constructed by forward induction from (5.1.9).

While the existence of ϵ-optimal controls is assured by Theorem 14.1 of Hinderer [8], its construction is not related to the solution of the dynamic programming equation.

5.2 Definitions and preliminary results

Throughout this chapter it will be assumed that Y is a sufficient statistic for the structure V (Definitions 3.2.2 and 3.2.6) and that the loss function has structure V (Definition 3.2.5). That is,

$$L = \sum_{\tau=1}^{T} c_t^V(V_t(x_t, u^{t-1}), u_{t-1}) \quad . \tag{5.2.1}$$

Definition 5.2.1. Conditional cost functions in Y are defined by

$$c_t^Y(y_t, u_{t-1}) = \int_{\mathcal{V}_t} c_t^V(v_t, u_{t-1}) G_t^{Y,V}(y_t, dv_t) \tag{5.2.2}$$

where $G_t^{Y,V}$ is the stochastic kernel of Definition 3.2.6.

Lemma 5.2.1. Let $c_{t+1}^Y(y_{t+1}, u_t)$ satisfy (5.2.2) and let $\ell_{t+1}(y_{t+1})$ be non-negative and measurable ($+\infty$ is admitted), and let K_t^Y be given by the definition of the sufficient statistic Y (Definition 3.2.2). Then $c_{t+1}^Y(y_{t+1}, u_t)$ and $L_t(y_t, u_t)$ defined by

$$L_t(y_t, u_t) = \int_{\mathcal{Y}_{t+1}} [\ell_{t+1}(y_{t+1}) + c_{t+1}^Y(y_{t+1}, u_t)] K_t^Y(y_t, u_t; dy_{t+1}) \tag{5.2.3}$$

are non-negative, measurable, and satisfy

$$E_{\{u\}}[c_{t+1}^V(V_{t+1}, u_t) | z^{t+1}] = c_{t+1}^Y(Y_{t+1}, u_t) \qquad \text{a.s. } P_{\{u\}}^{t+1} \tag{5.2.4}$$

and

$$E_{\{u\}}[c_{t+1}^V(V_{t+1}, u_t) + \ell_{t+1}(Y_{t+1}) | z^t] = L_t(Y_t, u_t) \qquad \text{a.s. } P_{\{u\}}^{t+1} \tag{5.2.5}$$

for all control laws $\{u\}$. (The notational convention (3.2.20) is followed.)

Proof: From Lemma A.1.1, the functions c_{t+1}^Y and L_{t+1} are non-negative and measurable. The equation (5.2.4) follows from the defining property (3.2.8) of $G_{t+1}^{Y,V}$ and Lemma A.1.5. From (5.2.4), the defining property

(3.2.2) of K_t^Y and Lemma A.1.5

$$E_{\{u\}}[c_{t+1}^V(V_{t+1},u_t) + \ell_{t+1}(Y_{t+1})|z^t]$$

$$= E_{\{u\}}[E_{\{u\}}[c_{t+1}^V(V_{t+1},u_t)|z^{t+1}]|z^t] + E_{\{u\}}[\ell_{t+1}(Y_{t+1})|z^t]$$

$$= E_{\{u\}}[c_{t+1}^Y(Y_{t+1},u_t) + \ell_{t+1}(Y_{t+1})|z^t]$$

$$= L_t(Y_t,u_t) \qquad \text{a.s. } P_{\{u\}}^t \ .$$

Lemma 5.2.2. Let $\ell_\tau(y_\tau)$ be non-negative and measurable, let $L_\tau(y_\tau,u_\tau)$ be defined by (5.2.3) for $\tau = s, s+1,\ldots,T-1$, let

$$L_T = \ell_T = 0 \ , \tag{5.2.6}$$

and let $t \leq s$.

i) If $\{u\}$ satisfies

$$E_{\{u\}}[\sum_{\alpha=1}^{\tau} c_\alpha^V + L_\tau(Y_\tau,u_\tau)|z^t] \leq E_{\{u\}}[\sum_{\alpha=1}^{\tau} c_\alpha^V + \ell_\tau(Y_\tau)|z^t] \qquad \text{a.s. } P_{\{u\}}^t \tag{5.2.7}$$

for $\tau = s+1, s+2,\ldots,T-1$, then

$$E_{\{u\}}[L|z^t] = E_{\{u\}}[\sum_{\alpha=1}^{s} c_\alpha^V + L_s(Y_s,u_s)|z^t] \qquad \text{a.s. } P_{\{u\}}^t \ . \tag{5.2.8}$$

ii) If $\{u\}$ satisfies

$$E_{\{u\}}[\sum_{\alpha=1}^{\tau} c_\alpha^V + L_\tau(Y_\tau,u_\tau)|z^t] = E_{\{u\}}[\sum_{\alpha=1}^{\tau} c_\alpha^V + \ell_\tau(Y_\tau)|z^t] \qquad \text{a.s. } P_{\{u\}}^t \tag{5.2.9}$$

for $\tau = s+1 , s+2,\ldots,T-1$, then

$$E_{\{u\}}[L|z^t] = E_{\{u\}}[\sum_{\alpha=1}^{s} c_\alpha^V + L_s(Y_s,u_s)|z^t] \qquad \text{a.s. } P_{\{u\}}^t \ . \tag{5.2.10}$$

iii) If $\{u\}$ satisfies

$$E_{\{u\}}[\sum_{\alpha=1}^{\tau} c_\alpha^V + L_\tau(Y_\tau,u_\tau)|z^t]$$

$$(5.2.11) \qquad \geq E_{\{u\}}[\sum_{\alpha=1}^{\tau} c_\alpha^V + \ell_\tau(Y_\tau)|z^t] - \epsilon_\tau \qquad \text{a.s. } P_{\{u\}}^t$$

for $\tau = s+1, s+2, \ldots, T-1$, then

$$(5.2.12) \qquad E_{\{u\}}[L|z^t] \geq E_{\{u\}}[\sum_{\alpha=1}^{s} c_\alpha^V + L_s(Y_s,u_s)|z^t] - \sum_{\alpha=s+1}^{T-1} \epsilon_\alpha \qquad \text{a.s. } P_{\{u\}}^t$$

where $0 \leq \epsilon_\tau < \infty$.

Proof: The proof is essentially the same for all three parts. Only iii) will be proved. For t fixed the proof is by backward induction on s . For $s = T$ and $L_T = 0$, (5.2.12) follows from (5.2.1). Assume then that (5.2.12) holds for $s+1$. From the induction hypothesis, (5.2.11) for $\tau = s+1$, the smoothing property of conditional expectations, and (5.2.5) of Lemma 5.2.1

$$\begin{aligned}
E_{\{u\}}[L|z^t] &\geq E_{\{u\}}[\sum_{\alpha=1}^{s+1} c_\alpha^V + L_{s+1}(Y_{s+1},u_{s+1})|z^t] - \sum_{\alpha=s+2}^{T-1} \epsilon_\alpha \\
&\geq E_{\{u\}}[\sum_{\alpha=1}^{s+1} c_\alpha^V + \ell_{s+1}(Y_{s+1})|z^t] - \epsilon_{s+1} - \sum_{\alpha=s+2}^{T} \epsilon_\alpha \\
&= E_{\{u\}}[\sum_{\alpha=1}^{s} c_\alpha^V + E_{\{u\}}[c_{s+1}^V + \ell_{s+1}(Y_{s+1})|z^s]|z^t] - \sum_{\alpha=s+1}^{T-1} \epsilon_\alpha \\
&= E_{\{u\}}[\sum_{\alpha=1}^{s} c_\alpha^V + L_s(Y_s,u_s)|z^t] - \sum_{\alpha=s+1}^{T-1} \epsilon_\alpha \qquad \text{a.s. } P_{\{u\}}^t .
\end{aligned}$$

Thus (5.2.12) holds for s .

<u>Definition</u> 5.2.2. Let $\{u\}$ be a control law and $\{U^Y\}$ a control law in the statistic Y_t (Definition 3.2.3) . Then the <u>compound law</u> $\{u^t\} \vee \{U^Y\} = \{u^{t-1}\} \vee u_t(\cdot) \vee \{U^Y\}$ has control functions $\tilde{u}_\tau(z^\tau)$ defined by

$$(5.2.13) \qquad \tilde{u}_\tau(z^\tau) = u_\tau(z^\tau) \qquad \tau = 0,1,\ldots,t$$

$$(5.2.14) \qquad \tilde{u}_\tau(z^\tau) = \hat{U}_\tau^Y(Y_\tau(z^\tau,\tilde{u}^{\tau-1}(z^{\tau-1}))) \qquad \tau = t+1,\ldots,T-1 .$$

Thus the control $\{u\}$ is used thru time t and after that $\{U^Y\}$ is used. When $\{U^Y\}$ is treated as a control law, it is defined by the control functions $\tilde{u}_\tau(z^\tau)$ which satisfy (5.2.14) where $t = 1$ and $\tilde{u}_0 = U_0$.

Lemma 5.2.3. Let L_T and ℓ_T satisfy (5.2.6), and for $\tau = t, t+1, \ldots, T-1$ let ℓ_τ, L_τ, and $\{U\}$ satisfy (5.2.3) and

$$\ell_\tau(y_\tau) = L_\tau(y_\tau, U_\tau(y_\tau)) . \tag{5.2.15}$$

Then for all $\{u\}$

$$E_{\{u^t\}\vee\{U\}}[L|z^t] = E_{\{u\}}[\sum_{\tau=1}^{t} c_\tau^V|z^t] + L_t(Y_t, u_t(z^t)) \qquad \text{a.s. } P_{\{u\}}^t \tag{5.2.16}$$

and

$$E_{\{u^{t-1}\}\vee\{U\}}[L|z^t] = E_{\{u\}}[\sum_{\tau=1}^{t} c_\tau^V|z^t] + \ell_t(Y_t) \qquad \text{a.s. } P_{\{u\}}^t . \tag{5.2.17}$$

Proof: From (5.2.15) the assumption of Lemma 5.2.2 ii) holds for s=t and all control laws. From (5.2.10) for the law $\{u^t\}\vee\{U\}$

$$E_{\{u^t\}\vee\{U\}}[L|z^t] = E_{\{u^t\}\vee\{U\}}[\sum_{\alpha=1}^{t} c_\alpha^V|z^t] + L_t(Y_t, u_t(z^t)) \quad \text{a.s. } P_{\{u\}}^t .$$

From (5.2.10) for $\{u^{t-1}\}\vee\{U\}$ and (5.2.15) for $\tau = t$

$$E_{\{u^{t-1}\}\vee\{U\}}[L|z^t] = E_{\{u^{t-1}\}\vee\{U\}}[\sum_{\alpha=1}^{t} c_\alpha^V|z^t] + L_t(Y_t, U_t(Y_t))$$

$$= E_{\{u^{t-1}\}\vee\{U\}}[\sum_{\alpha=1}^{t} c_\alpha^V|z^t] + \ell_t(Y_t) \qquad \text{a.s. } P_{\{u\}}^t .$$

Since the joint distribution of $\sum_{\alpha=1}^{t} c_\alpha^V$ and z^t depends only on the truncated law $\{u^{t-1}\}$

$$E_{\{u^t\}\vee\{U\}}[\sum_{\alpha=1}^{t} c_\alpha^V|z^t] = E_{\{u^{t-1}\}\vee\{U\}}[\sum_{\alpha=1}^{t} c_\alpha^V|z^t]$$

$$= E_{\{u\}}[\sum_{\alpha=1}^{t} c_\alpha^V|z^t] .$$

The equations (5.2.16) and (5.2.17) follow.

The definition of a selection class is motivated by Theorem 5.4.8 in which it is shown that under "natural" assumptions the classes $\mathcal{D}_\rho$ (Definition 3.2.10) are selection classes.

The following notation will be required in the definition of a selection class. For

(5.2.18) $\underline{\varepsilon} = (\varepsilon_0, \varepsilon_1, \ldots, \varepsilon_{T-1})$

where

(5.2.19) $0 < \varepsilon_t < \infty \qquad t = 0,1,\ldots,T-1$

let

(5.2.20) $\underline{\varepsilon}_t = (\varepsilon_t, \varepsilon_{t+1}, \ldots, \varepsilon_{T-1})$

<u>Definition</u> 5.2.3. The class $\mathcal{D}_0$ of control laws is a <u>selection class for</u> Y provided for each vector $\underline{\varepsilon}$ there exist a control law $\{\hat{U}^{Y,\mathcal{D}_0,\underline{\varepsilon}}\}$ in Y and non-negative measurable functions $L_t^{Y,\mathcal{D}_0,\underline{\varepsilon}_{t+1}}(y_t,u_t)$, $\tilde{\ell}_t^{Y,\mathcal{D}_0,\underline{\varepsilon}_{t+1}}(y_t)$, and $\ell_t^{Y,\mathcal{D}_0,\underline{\varepsilon}_t}(y_t)$ which satisfy for all laws $\{u\} \in \mathcal{D}_0$

(5.2.21) $\{u^{t-1}\} \vee \{\hat{U}^{Y,\mathcal{D}_0,\underline{\varepsilon}}\} \in \mathcal{D} \qquad t = 0,1,\ldots,T-1$

(5.2.22) $L^{Y,\mathcal{D}_0} = \ell_T^{Y,\mathcal{D}_0} = 0$

(5.2.23) $L_t^{Y,\mathcal{D}_0,\underline{\varepsilon}_{t+1}}(y_t,u_t) = \int_{\mathcal{Y}_{t+1}} [\ell^{Y,\mathcal{D}_0,\underline{\varepsilon}_{t+1}}(y_{t+1}) + c_{t+1}^Y(y_{t+1},u_t)]K_t^Y(y_t,u_t;dy_{t+1})$

(5.2.24) $\tilde{\ell}_t^{Y,\mathcal{D}_0,\underline{\varepsilon}_{t+1}}(y_t) = \inf_{u_t \in \mathcal{U}_t} L_t^{Y,\mathcal{D}_0,\underline{\varepsilon}_{t+1}}(y_t,u_t) \qquad \text{a.s. } (Y^{-1}P)_{\{u\}}^t$

(5.2.25) $\chi_{B_t}(y_t)\, L_t^{Y,\mathcal{D}_0,\underline{\varepsilon}_{t+1}}(y_t,\hat{U}_t^{Y,\mathcal{D}_0,\underline{\varepsilon}_t}(y_t)) = \chi_{B_t}(y_t)\tilde{\ell}_t^{Y,\mathcal{D}_0,\underline{\varepsilon}_{t+1}}(y_t) \text{ a.s.} (Y^{-1}P)_{\{u\}}^t$

(5.2.26)
$$\chi_{B_t^c}(y_t)\, L_t^{Y,\mathcal{D}_0,\underline{\varepsilon}_{t+1}}(y_t,\hat{U}_t^{Y,\mathcal{D}_0,\underline{\varepsilon}_t}(y_t))$$
$$\leq \chi_{B_t^c}(y_t)\tilde{\ell}_t^{Y,\mathcal{D}_0,\underline{\varepsilon}_{t+1}}(y_t) + \varepsilon_t \qquad \text{a.s. } (Y^{-1}P)_{\{u\}}^t$$

(5.2.27) $\ell^{Y,\mathcal{D}_0,\underline{\varepsilon}_t}(y_t) = L^{Y,\mathcal{D}_0,\underline{\varepsilon}_{t+1}}(y_t,\hat{U}_t^{Y,\mathcal{D}_0,\underline{\varepsilon}_t}(y_t))$

where $B_t \in \mathcal{B}_{\mathcal{Y}_t}$ and satisfies

(5.2.28) $B_t = \{y_t | \exists u_t^* \in \mathcal{U}_t \ni L_t^{Y,\mathcal{D}_0,\underline{\varepsilon}_{t+1}}(y_t,u_t^*) = \tilde{\ell}_t^{Y,\mathcal{D}_0,\underline{\varepsilon}_{t+1}}(y_t)\} \text{ a.s. } (Y^{-1}P)_{\{u\}}^t$

$t = 0,1,\ldots,T-1$. Further, it will be assumed that for

(5.2.29) $\underline{\varepsilon}^n = (\frac{1}{n}, \frac{1}{n}, \ldots, \frac{1}{n})$

the functions $\ell^{Y,\mathcal{D}_0,\underline{\varepsilon}^n}(y_t)$ and $L_t^{Y,\mathcal{D}_0,\underline{\varepsilon}^n}(y_t,u_t)$ are monotone non-increasing in n.

The sets B_t, of course, also depend on Y, $\mathcal{D}_0$, and $\underline{\varepsilon}_{t+1}$. When the intention is clear, the superscripts Y and $\mathcal{D}_0$ will be omitted and $\underline{\varepsilon}_t$ will be replaced by $\underline{\varepsilon}$ or omitted.

Lemma 5.2.4. For a selection class $\mathcal{D}_0$, the functions $\hat{\ell}^{Y,\mathcal{D}_0}(y_t)$ and $\hat{L}_t^{Y,\mathcal{D}_0}(y_t,u_t)$ defined by

$$(5.2.30)\qquad \ell_t^{Y,\mathcal{D}_0,\underline{\varepsilon}^n}(y_t) \downarrow \hat{\ell}_t^{Y,\mathcal{D}_0}(y_t)$$

$$(5.2.31)\qquad L_t^{Y,\mathcal{D}_0,\underline{\varepsilon}^n}(y_t) \downarrow \hat{L}_t^{Y,\mathcal{D}_0}(y_t,u_t)$$

are non-negative, measurable, satisfy (5.2.3), (5.2.6), and

$$(5.2.32)\qquad \hat{\ell}_t^{Y,\mathcal{D}_0}(y_t) = \inf_{u_t \in \mathcal{U}_t} \hat{L}_t^{Y,\mathcal{D}_0}(y_t,u_t) \qquad \text{a.s. } (Y^{-1}P)^t_{\{u\}}$$

for $\{u\} \in \mathcal{D}_0$.

Proof: Since by assumption the functions (5.2.30) and (5.2.31) are non-negative, measurable, and monotone non-increasing in n, they converge to limits which are non-negative and measurable. The relation (5.3.2) is easily deduced from (5.2.23) for $\underline{\varepsilon}^n$, (5.2.30), (5.2.31), and the monotone convergence theorem.

From (5.2.24)-(5.2.27), and (5.2.29)

$$(5.2.33)\quad \inf_{u_t \in \mathcal{U}_t} L_t^{\underline{\varepsilon}^n}(y_t,u_t) \le \ell_t^{\underline{\varepsilon}^n}(y_t) \le \inf_{u_t \in \mathcal{U}_t} L_t^{\underline{\varepsilon}^n}(y_t,u_t) + \frac{1}{n} \qquad \text{a.s. } (Y^{-1}P)^t_{\{u\}} \quad .$$

From the monotoneity of $L_t^{\underline{\varepsilon}^n}$

$$\lim_{n\to\infty} \inf_{u_t \in \mathcal{U}_t} L_t^{\underline{\varepsilon}^n}(y_t,u_t) = \inf_{u_t \in \mathcal{U}_t} \hat{L}_t(y_t,u_t)$$

for all y_t. Thus (5.2.32) holds except for y_t in the union of the exceptional sets of (5.2.33) for $n = 1,2,\ldots$.

5.3 Properties of selection classes

From (5.2.23), (5.2.22), and (5.2.27) of the definition of a selection class, the relations (5.2.3), (5.2.6), and (5.2.15) hold for $\ell_t^{Y,\mathcal{D}_0,\underline{\varepsilon}}$, $L_t^{Y,\mathcal{D}_0,\underline{\varepsilon}}$, and $\{\hat{U}^{Y,\mathcal{D}_0,\underline{\varepsilon}}\}$, $t = 0,1,2,\ldots,T-1$. Thus the results of Lemmas 5.2.1, 5.2.2, and 5.2.3 hold for these functions and all admissible control laws $\{u\} \in \mathcal{D}$. Further properties of these functions and the selection class $\mathcal{D}_0$ will be developed in a series of lemmas.

Throughout this section a fixed sufficient statistic Y and selection class $\mathcal{D}_0$ is considered. Thus these superscripts will be omitted.

<u>Lemma</u> 5.3.1. Let $\mathcal{D}_0$ be a selection class. Then for $\{u\} \in \mathcal{D}_0$

$$\tilde{\ell}_t^{\underline{\varepsilon}}(Y_t) \leq L_t^{\underline{\varepsilon}}(Y_t,\hat{U}_t^{\underline{\varepsilon}}(Y_t)) = \ell_t^{\underline{\varepsilon}}(Y_t) \qquad \text{a.s. } P_{\{u\}}^t \tag{5.3.1}$$

$$\tilde{\ell}_t^{\underline{\varepsilon}}(Y_t) \leq L_t^{\underline{\varepsilon}}(Y_t,u_t(z^t)) \qquad \text{a.s. } P_{\{u\}}^t \tag{5.3.2}$$

$$L_t^{\underline{\varepsilon}}(Y_t,\hat{U}_t^{\underline{\varepsilon}}(Y_t)) = \ell_t^{\underline{\varepsilon}}(Y_t) \leq \tilde{\ell}_t^{\underline{\varepsilon}}(Y_t) + \varepsilon_t \qquad \text{a.s. } P_{\{u\}}^t \tag{5.3.3}$$

$$L_t^{\underline{\varepsilon}}(Y_t,\hat{U}_t^{\underline{\varepsilon}}(Y_t)) = \ell_t^{\underline{\varepsilon}}(Y_t) \leq L_t^{\underline{\varepsilon}}(Y_t,u_t(z^t)) + \varepsilon_t \qquad \text{a.s. } P_{\{u\}}^t \; . \tag{5.3.4}$$

Proof: For all z^t

$$\inf_{u_t \in \mathcal{U}_t} L_t^{\underline{\varepsilon}}(Y_t,u_t) \leq L_t^{\underline{\varepsilon}}(Y_t,u_t(z^t))$$

and

$$\inf_{u_t \in \mathcal{U}_t} L_t^{\underline{\varepsilon}}(Y_t,u_t) \leq L_t^{\underline{\varepsilon}}(Y_t,\hat{U}_t^{\underline{\varepsilon}}(Y_t)) \; .$$

Thus (5.3.1) and (5.3.2) follow from (5.2.24) and the definition of $(Y^{-1}P)_{\{u\}}^t$ (Definition 3.2.9). Adding (5.2.25) and (5.2.26)

$$L_t^{\underline{\varepsilon}}(y_t,\hat{U}_t^{\underline{\varepsilon}}(y_t)) \leq \tilde{\ell}_t^{\underline{\varepsilon}}(y_t) + \varepsilon_t \qquad \text{a.s. } (Y^{-1}P)_{\{u\}}^t \; .$$

Thus (5.3.3) follows from (5.2.27) and the definition of $(Y^{-1}P)_{\{u\}}^t$. The result (5.3.4) follows from (5.3.2) and (5.3.3).

Lemma 5.3.2. Let $\mathcal{D}_0$ be a selection class, and let $\{u\} \in \mathcal{D}_0$, $B^* \in \mathcal{B}_Z^t$, and ε_{t+1} satisfy

$$(5.3.5) \qquad \chi_{B^*}(z^t)\, E[L_s^{\varepsilon_{s+1}}(Y_s,u_s(z^s))\,|\,z^t] = \chi_{B^*}(z^t)\, E[\tilde{\ell}_s^{\varepsilon_{s+1}}(Y_s)\,|\,z^t] < \infty \qquad \text{a.s. } P^t_{\{u\}}$$

where $t \leq s$. Then

$$(5.3.6) \quad \chi_{B^*}(z^t)E[L_s^{\varepsilon_{s+1}}(Y_s,u_s(z^s))\,|\,z^t] = \chi_{B^*}(z^t)E[\ell_s^{\varepsilon_s}(Y_s)\,|\,z^t] \qquad \text{a.s. } P^t_{\{u\}}$$

for all $\varepsilon_t > 0$.

Proof: From (5.3.2) of Lemma 5.3.1 and the finiteness assumption of (5.3.5)

$$0 \leq \chi_{B^*}(z^t)\tilde{\ell}_s^{\varepsilon}(Y_s) \leq \chi_{B^*}(z^t)L_s^{\varepsilon}(Y_s,u_s(z^s)) < \infty \qquad \text{a.s. } P^s_{\{u\}} .$$

Thus from (5.3.5)

$$E_{\{u\}}[\,|\chi_{B^*}L_s^{\varepsilon} - \chi_{B^*}\tilde{\ell}_s^{\varepsilon}|\,] = E_{\{u\}}[\chi_{B^*}(L_s^{\varepsilon} - \tilde{\ell}_s^{\varepsilon})]$$
$$= E_{\{u\}}[\chi_{B^*}E_{\{u\}}[L_s^{\varepsilon} - \tilde{\ell}_s^{\varepsilon}\,|\,z^t]] = 0 .$$

Thus

$$\chi_{B^*}(z^t)L_s^{\varepsilon}(Y_s,u_s(z^s)) = \chi_{B^*}(z^t)\tilde{\ell}_s^{\varepsilon}(Y_s) < \infty \qquad \text{a.s. } P^s_{\{u\}} .$$

Let B_s be defined by (5.2.28) and $u_s^* = u_s(z^s)$. Then $z^t \in B^*$ implies that

$$Y_s(z^s, u^{s-1}(z^{s-1})) \in B_s \qquad \text{a.s. } P^s_{\{u\}} .$$

Thus from (5.2.27) and (5.2.25),

$$\chi_{B^*}(z^t)\ell_s^{\varepsilon}(Y_s) = \chi_{B^*}(z^t)L_s^{\varepsilon}(Y_s,\hat{U}_s^{\varepsilon}(Y_s))$$
$$= \chi_{B^*}(z^t)\tilde{\ell}_s^{\varepsilon}(Y_s) = \chi_{B^*}(z^t)L_s^{\varepsilon}(Y_s,u_s(z^s)) \qquad \text{a.s. } P^s_{\{u\}} .$$

Taking conditional expectations, (5.3.6) follows.

Theorem 5.3.1. For $\{u\} \in \mathcal{D}_0$, a selection class, and all $\underline{\epsilon}$,

$$\{u^{t-1}\} \vee \{\hat{U}^{\underline{\epsilon}}\} \in \mathcal{D} \tag{5.3.7}$$

and

$$\begin{aligned} E_{\{u^{t-1}\} \vee \{\hat{U}^{\underline{\epsilon}}\}}[L|z^t] &= E_{\{u\}}[\sum_{\tau=1}^{t} c_\tau^V|z^t] + \ell_t^{\underline{\epsilon}t}(Y_t) \\ &\leq E_{\{u\}}[L|z^t] + \sum_{\tau=t}^{T-1} \epsilon_\tau \qquad \text{a.s. } P_{\{u\}}^t \end{aligned} \tag{5.3.8}$$

for all $0 \leq t \leq T$.

Proof: From (5.3.4) of Lemma 5.3.1, the assumptions of Lemma 5.2.2 iii) are fulfilled for $L_t^{\underline{\epsilon}}$, $\ell_t^{\underline{\epsilon}}$, $\{u\}$, and $s = t$. Thus from (5.2.12), (5.3.4) of Lemma 5.3.1, and (5.2.17) of Lemma 5.2.3

$$\begin{aligned} E_{\{u\}}[L|z^t] &\geq E_{\{u\}}[\sum_{\tau=1}^{t} c_\tau^V|z^t] + L_t^{\underline{\epsilon}}(Y_t,u_t(z^t)) - \sum_{\tau=t+1}^{T-1} \epsilon_\tau \\ &\geq E_{\{u\}}[\sum_{\tau=1}^{t} c_\tau^V|z^t] + \ell_t^{\underline{\epsilon}}(Y_t) - \epsilon_t - \sum_{\tau=t+1}^{T-1} \epsilon_\tau \\ &= E_{\{u^{t-1}\} \vee \{\hat{U}^{\underline{\epsilon}}\}}[L|z^t] - \sum_{\tau=t}^{T-1} \epsilon_\tau \qquad \text{a.s. } P_{\{u\}}^t . \end{aligned}$$

The result (5.3.7) is just a restatement of (5.2.21) of Definition 5.2.3.

Lemma 5.3.3. Let $\{\hat{u}\} \in \mathcal{D}_0$, a selection class, satisfy

$$E_{\{\hat{u}\}}[L|z^t] = E_{\{\hat{u}\}}[\hat{F}_\tau(z^\tau,\{\hat{u}\})|z^t] \qquad \text{a.s. } P_{\{\hat{u}\}}^t \tag{5.3.9}$$

for $\tau = t,t+1,\ldots,T$. Then for all $\underline{\epsilon}$

$$\begin{aligned} &E_{\{\hat{u}\}}[\sum_{\alpha=1}^{\tau} c_\alpha^V + L_\tau^{\underline{\epsilon}}(\hat{Y}_\tau,\hat{u}_\tau(z^\tau))|z^t] \\ &= E_{\{\hat{u}\}}[\sum_{\alpha=1}^{\tau} c_\alpha^V + \ell_\tau^{\underline{\epsilon}}(\hat{Y}_\tau)|z^t] \qquad \text{a.s. } P_{\{\hat{u}\}}^t \end{aligned} \tag{5.3.10}$$

for $\tau = t,t+1,\ldots,T$.

Proof: For t fixed, the proof is by backward induction on τ. For $\tau = T$, from (5.2.22), (5.3.10) is trivial. Assume then that (5.3.10) holds for $\tau = s+1,s+2,\ldots,T$ where $t \leq s$. From the induction hypothesis, the assumptions of Lemma 5.2.2 ii) are fulfilled. Thus, from that lemma

(5.3.11) $$E_{\{\hat{u}\}}[L|z^t] = E_{\{\hat{u}\}}[\sum_{\alpha=1}^{s} c_\alpha^V + L_s^{\varepsilon}(\hat{Y}_s,\hat{u}_s)|z^t] \quad \text{a.s. } P_{\{\hat{u}\}}^t \quad .$$

From (5.3.11), (5.3.9) for $\tau = s$, the definition of $\hat{F}_s(z^s,\{\hat{u}\})$, (5.2.17) of Lemma 5.2.3 for $t = s$, and (5.3.3) of Lemma 5.3.1

$$E_{\{\hat{u}\}}[\sum_{\alpha=1}^{s} c_\alpha^V + L_s^{\varepsilon}(\hat{Y}_s,\hat{u}_s)|z^t] = E_{\{\hat{u}\}}[\hat{F}_s(z^s,\{\hat{u}\})|z^t]$$

(5.3.12) $$\leq E_{\{\hat{u}\}}[E^{\varepsilon}_{\{\hat{u}^{s-1}\}\vee\{\hat{U}\}}[L|z^s]|z^t]$$

$$= E_{\{\hat{u}\}}[E_{\{\hat{u}\}}[\sum_{\alpha=1}^{s} c_\alpha^V|z^s] + \ell_t^{\varepsilon}(\hat{Y}_s)|z^t]$$

$$\leq E_{\{\hat{u}\}}[\sum_{\alpha=1}^{s} c_\alpha^V + \tilde{\ell}_s^{\varepsilon}(\hat{Y}_s)|z^t] + \epsilon_s \quad \text{a.s. } P_{\{u\}}^t \quad .$$

From (5.3.2) of Lemma 5.3.1 and (5.3.12) for all $\epsilon_s > 0$

$$E_{\{\hat{u}\}}[\sum_{\alpha=1}^{s} c_\alpha^V + \tilde{\ell}_s^{\varepsilon s+1}(\hat{Y}_s)|z^t] \leq E_{\{\hat{u}\}}[\sum_{\alpha=1}^{s} c_\alpha^V + L_s^{\varepsilon s+1}(\hat{Y}_s,\hat{u}_s(z^s))|z^t]$$

$$\leq E_{\{\hat{u}\}}[\sum_{\alpha=1}^{s} c_\alpha^V + \tilde{\ell}_s^{\varepsilon s+1}(\hat{Y}_s)|z^t] + \epsilon_s \quad \text{a.s. } P_{\{\hat{u}\}}^t \quad .$$

It follows that

$$E_{\{\hat{u}\}}[\sum_{\alpha=1}^{s} c_\alpha^V + \tilde{\ell}_s^{\varepsilon s+1}(\hat{Y}_s)|z^t]$$

$$= E_{\{\hat{u}\}}[\sum_{\alpha=1}^{s} c_\alpha^V + L_s^{\varepsilon s+1}(\hat{Y}_s,\hat{u}_s(z^s))|z^t] \quad \text{a.s. } P_{\{\hat{u}\}}^t \quad .$$

Let B^* be the set on which

$$E_{\{\hat{u}\}}[\sum_{\alpha=1}^{s} c_\alpha^V|z^t] < \infty \quad \text{and} \quad E_{\{\hat{u}\}}[\tilde{\ell}_s^{\varepsilon}(\hat{Y}_s)|z^t] < \infty \quad .$$

Then from Lemma 5.3.2 on B^*

$$E_{\{\hat{u}\}}[L_s^{\varepsilon}(\hat{Y}_s,\hat{u}_s)|z^t] = E_{\{\hat{u}\}}[\ell_s^{\varepsilon}(\hat{Y}_s)|z^t] \quad \text{a.s. } P_{\{\hat{u}\}}^t$$

and (5.3.10) holds on B^* for $\tau = s$. From (5.3.2) of Lemma 5.3.1, (5.3.10) also holds on

$$E_{\{\hat{u}\}}[L_s^{\varepsilon}(\hat{Y}_s,\hat{u}_s)|z^t] \geq E_{\{\hat{u}\}}[\tilde{\ell}_s^{\varepsilon}(\hat{Y}_s)|z^t] = \infty \quad .$$

On

$$E_{\{\hat{u}\}}[\sum_{\alpha=1}^{s} c_\alpha^V | z^t] = \infty ,$$

(5.3.10) obviously holds. Thus (5.3.10) holds a.s. $P^t_{\{\hat{u}\}}$.

Theorem 5.3.2. Let $\{\hat{u}\} \in \mathcal{D}_0$, a selection class, satisfy

$$(5.3.13) \qquad E_{\{\hat{u}\}}[\hat{F}_\tau(z^\tau,\{\hat{u}\})|z^t] = E_{\{\hat{u}\}}[L|z^t] \qquad \text{a.s. } P^t_{\{\hat{u}\}}$$

for $\tau = t,t+1,\ldots,T$. Then (5.3.7) holds and

$$(5.3.14) \qquad E_{\{\hat{u}\}}[L|z^t] = E_{\{\hat{u}^{t-1}\}\vee\{\hat{U}^{\underline{\varepsilon}}\}}[L|z^t] \qquad \text{a.s. } P^t_{\{u\}}$$

for all $\underline{\varepsilon}$.

Proof: From Lemma 5.3.3 the assumptions of Lemma 5.2.2 ii) are satisfied for $\boldsymbol{\ell}_\tau^{\underline{\varepsilon}}$, $L_\tau^{\underline{\varepsilon}}$, $\{\hat{u}\}$ and $s = t$. Thus from (5.2.10), (5.3.10) for $\tau = t$, and (5.2.17) of Lemma 5.2.3

$$E_{\{\hat{u}\}}[L|z^t] = E_{\{\hat{u}\}}[\sum_{\alpha=1}^{t} c_\alpha^V|z^t] + L_t^{\underline{\varepsilon}}(\hat{Y}_t,\hat{u}_t(z^t))$$

$$= E_{\{\hat{u}\}}[\sum_{\alpha=1}^{t} c_\alpha^V|z^t] + \boldsymbol{\ell}_t^{\underline{\varepsilon}}(\hat{Y}_t)$$

$$= E_{\{\hat{u}^{t-1}\}\vee\{\hat{U}^{\underline{\varepsilon}}\}}[L|z^t] \qquad \text{a.s. } P^t_{\{\hat{u}\}} .$$

Lemma 5.3.4. Let $\mathcal{D}_0$ be a selection class and let $\{u\} \in \mathcal{D}$ satisfy

$$(5.3.15) \qquad \{u^{t-1}\} \vee \{\hat{U}^{\underline{\varepsilon}^n}\} \in \mathcal{D} \qquad n = 1,2,\ldots .$$

Then

$$(5.3.16) \qquad \hat{F}_t(z^t,\{u\}) \leq \hat{\boldsymbol{\ell}}_t(Y_t) + E_{\{u\}}[\sum_{\tau=1}^{t} c_\tau^V|z^t] \qquad \text{a.s. } P^t_{\{u\}} ,$$

and for $\{u\} \in \mathcal{D}_0$

$$(5.3.17) \qquad \hat{\boldsymbol{\ell}}_t(Y_t) + E_{\{u\}}[\sum_{\tau=1}^{t} c_\tau^V|z^t] \leq E_{\{u\}}[L|z^t] \qquad \text{a.s. } P^t_{\{u\}} .$$

Proof: From the definition of $\hat{F}_t$, (5.3.15), and (5.2.17) of Lemma 5.2.3

$$\hat{F}_t(z^t,\{u\}) \leq E_{\{u^{t-1}\}\vee\{\hat{U}^{\epsilon^n}\}}[L|z^t] = \ell_t^{\epsilon^n}(Y_t) + E_{\{u\}}[\sum_{\tau=1}^{t} c_\tau^V|z^t] \quad \text{a.s. } P^t_{\{u\}} .$$

The result (5.3.16) then follows from the definition (5.2.30) of $\hat{\ell}_t$.

For $\{u\} \in \mathcal{D}_0$, from (5.2.17) of Lemma 5.2.3, (5.3.8) of Theorem 5.3.1, and the definition (5.2.29) of $\underline{\epsilon}^n$

$$\ell_t^{\epsilon^n}(Y_t) + E_{\{u\}}[\sum_{\tau=1}^{t} c_\tau^V|z^t] = E_{\{u^{t-1}\}\vee\{\hat{U}^{\epsilon^n}\}}[L|z^t]$$

$$\leq E_{\{u\}}[L|z^t] + \sum_{\tau=t}^{T-1} \epsilon_\tau^n = E_{\{u\}}[L|z^t] + \frac{T-t}{n} \quad \text{a.s. } P^t_{\{u\}} .$$

The result (5.3.17) then follows from (5.2.30) by letting $n \to \infty$.

<u>Lemma</u> 5.3.5. Let $\mathcal{D}_0$ be a selection class and let $\{\hat{u}\} \in \mathcal{D}_0$ satisfy

$$(5.3.18) \qquad \begin{aligned} &E_{\{\hat{u}\}}[\sum_{\alpha=1}^{\tau} c_\alpha^V|z^\tau] + L_\tau^{\epsilon}(\hat{Y}_\tau,\hat{u}_\tau(z^\tau)) \\ &\leq E_{\{\hat{u}\}}[\sum_{\alpha=1}^{\tau} c_\alpha^V|z^\tau] + \ell_\tau^{\epsilon}(\hat{Y}_\tau) \qquad \text{a.s. } P^\tau_{\{\hat{u}\}} \end{aligned}$$

for $\tau=t,t+1,\ldots,T-1$. Then for all $\{u\} \in \mathcal{D}_0$ such that

$$(5.3.19) \qquad \{u^{t-1}\} = \{\hat{u}^{t-1}\} \quad ,$$

$$(5.3.20) \qquad \begin{aligned} &E_{\{\hat{u}\}}[L|z^t] \leq E_{\{\hat{u}^{t-1}\}\vee\{\hat{U}^{\epsilon}\}}[L|z^t] \\ &\leq E_{\{u\}}[L|z^t] + \sum_{\tau=t}^{T-1} \epsilon_\tau \qquad \text{a.s. } P^t_{\{u\}} . \end{aligned}$$

Proof: From (5.3.18) the assumptions of Lemma 5.2.2 i) are fulfilled for ℓ_τ^{ϵ} , L_τ^{ϵ} , $\{\hat{u}\}$, and $s=t$. Thus from (5.2.8), (5.3.18) for $s=t$, (5.2.17) of Lemma 5.2.3, (5.3.19), and (5.3.8) of Theorem 5.3.1

$$E_{\{\hat{u}\}}[L|z^t] \leq E_{\{\hat{u}\}}[\sum_{\alpha=1}^{t} c_\alpha^V|z^t] + L_t^{\epsilon}(\hat{Y}_t,\hat{u}_t(z^t))$$

$$\leq E_{\{\hat{u}\}}[\sum_{\alpha=1}^{t} c_\alpha^V|z^t] + \ell_t^{\epsilon}(\hat{Y}_t) = E_{\{\hat{u}^{t-1}\}\vee\{\hat{U}^{\epsilon}\}}[L|z^t]$$

$$= E_{\{u^{t-1}\}\vee\{\hat{U}^{\epsilon}\}}[L|z^t] \leq E_{\{u\}}[L|z^t] + \sum_{\tau=t}^{T-1} \epsilon_\tau \qquad \text{a.s. } P^t_{\{u\}} .$$

Lemma 5.3.6. Let $\mathcal{D}_0$ be a selection class and let $\{\hat{u}\} \in \mathcal{D}_0$ satisfy

$$E_{\{\hat{u}\}}\left[\sum_{\alpha=1}^{\tau} c_\alpha^V + \hat{L}_\tau(\hat{Y}_\tau,\hat{u}_\tau(z^\tau))\,|\,z^t\right]$$

(5.3.21)
$$= E_{\{\hat{u}\}}\left[\sum_{\alpha=1}^{\tau} c_\alpha^V + \hat{\ell}_\tau(\hat{Y}_\tau)\,|\,z^t\right] \qquad \text{a.s. } P^t_{\{\hat{u}\}}$$

for $\tau=t,t+1.,,,T-1$. Then for all $\{u\} \in \mathcal{D}_0$ such that (5.3.19) holds,

(5.3.22)
$$E_{\{\hat{u}\}}[L|z^t] \leq E_{\{u\}}[L|z^t] \qquad \text{a.s. } P^t_{\{u\}} \quad .$$

Proof: From Lemma 5.2.4, $\hat{\ell}_\tau$ and $\hat{L}_\tau$ satisfy (5.2.3) and (5.2.6). Thus the assumptions of Lemma 5.2.2 ii) are fulfilled for $\hat{\ell}_\tau$, $\hat{L}_\tau$, $\{\hat{u}\}$ and $s=t$. From (5.2.10), (5.3.21) for $\tau=t$, (5.3.19) and (5.3.17) of Lemma 5.3.4

$$E_{\{\hat{u}\}}[L|z^t] = E_{\{\hat{u}\}}\left[\sum_{\alpha=1}^{t} c_\alpha^V|z^t\right] + \hat{L}_t(\hat{Y}_t,\hat{u}_t(z^t))$$

$$= E_{\{\hat{u}^{t-1}\}}\left[\sum_{\alpha=1}^{t} c_\alpha^V|z^t\right] + \hat{\ell}_t(\hat{Y}_t) = E_{\{u\}}\left[\sum_{\alpha=1}^{t} c_\alpha^V|z^t\right] + \hat{\ell}_t(Y_t)$$

$$\leq E_{\{u\}}[L|z^t] \qquad \text{a.s. } P^t_{\{u\}} \quad .$$

5.4 Complete families of selection classes

Definition 5.4.1. The system has a finitely complete family of selection classes in Y (more simply, **finitely** complete selection classes in Y) provided for any pair of admissible laws $\{\bar{u}\} \in \mathcal{D}$ and $\{\bar{\bar{u}}\} \in \mathcal{D}$, there exists a selection class $\mathcal{D}_0$ that contains them both, $\{\bar{u}\} \in \mathcal{D}_0$, $\{\bar{\bar{u}}\} \in \mathcal{D}$.

Definition 5.4.2. The system has a countably complete family of selection classes in Y (countably complete selection classes in Y) provided for any countable class of admissible laws, $\bar{\mathcal{D}} \subset \mathcal{D}$, there exists a selection class $\mathcal{D}_0$ which contains $\bar{\mathcal{D}}$, $\bar{\mathcal{D}} \subset \mathcal{D}_0$.

Theorem 5.4.1 i) If the system has finitely complete selection classes in Y , then the class $\mathcal{D}$ has the finite ϵ-lattice property.

ii) If the system has countably complete selection classes, then $\mathcal{D}$ has the countable ϵ-lattice property.

Proof: The proof is the same in each case. Let $\bar{\mathcal{D}} \subset \mathcal{D}$ be finite (countable) and satisfy

$$(5.4.1) \qquad \{\bar{u}^{t-1}\} = \{\bar{\bar{u}}^{t-1}\}$$

for all $\{\bar{u}\} \in \bar{\mathcal{D}}$, $\{\bar{\bar{u}}\} \in \bar{\mathcal{D}}$. From Definition 5.4.1 (5.4.2) there exists a selection class $\mathcal{D}_0$ such that $\bar{\mathcal{D}} \subset \mathcal{D}_0$. For $\epsilon > 0$ take $\underline{\epsilon}$ such that

$$\sum_{\tau=t}^{T-1} \epsilon_\tau \leq \epsilon .$$

Then from (5.3.8) of Theorem 5.3.1

$$\{\bar{u}^{t-1}\} \vee \{\hat{U}^{Y,\mathcal{D}_0,\underline{\epsilon}}\} \in \mathcal{D}$$

and

$$\mathbf{E}_{\{\bar{u}^{t-1}\} \vee \{\hat{U}^{Y,\mathcal{D}_0,\underline{\epsilon}}\}} [L|z^t] \leq E_{\{\bar{u}\}} [L|z^t] + \epsilon \qquad \text{a.s. } P^t_{\{\bar{u}\}}$$

for all $\{\bar{u}\} \in \bar{\mathcal{D}}$. From (5.4.1)

$$\{\bar{u}^{t-1}\} \vee \{\hat{U}^{Y,\mathcal{D}_0,\underline{\epsilon}}\} = \{\tilde{u}\}$$

is the same law for all $\{\bar{u}\} \in \bar{\mathcal{D}}$, and clearly

$$\{\tilde{u}^{t-1}\} = \{\bar{u}^{t-1}\} .$$

Thus (4.2.11) and (4.2.12) hold and the result follows from Definition 4.2.5 (4.2.6).

<u>Theorem</u> 5.4.2. If the system has finitely complete selection classes in Y, then the class of controls in Y and the compound laws $\{u^{t-1}\} \vee \{U^Y\}$ enjoy the following completeness properties.

i) For $\{u\} \in \mathcal{D}$ and $\epsilon > 0$, there exists $\{\hat{U}^\epsilon\}$ such that $\{u^{t-1}\} \vee \{\hat{U}^\epsilon\} \in \mathcal{D}$ and

$$(5.4.2) \qquad E_{\{u^{t-1}\} \vee \{\hat{U}^{\epsilon}\}}[L|z^t] \leq E_{\{u\}}[L|z^t] + \epsilon \qquad \text{a.s. } P^t_{\{u\}}$$

for all t .

ii) For $\{u\} \in \mathcal{D}$ and t fixed, suppose that for all $\epsilon > 0$ there exists $\{\hat{u}^{\epsilon}\}$ ϵ-optimal for $\{u\}$ at t . Then for all $\epsilon > 0$ there exists $\{\hat{U}^{\epsilon}\}$ such that $\{u^{t-1}\} \vee \{\hat{U}^{\epsilon}\}$ is ϵ-optimal for $\{u\}$ at t .

iii) For $\epsilon > 0$, there exists $\{\hat{U}^{\epsilon}\}$, ϵ-optimal at $t = 0$.

Proof: (i) Let $\{\hat{U}^{\epsilon}\} = \{\hat{U}^{Y,\mathcal{D}_0,\underline{\epsilon}}\}$ where $\mathcal{D}_0$ is a selection class which contains $\{u\}$ and

$$\epsilon = \sum_{t=0}^{T-1} \epsilon_{\tau} \quad .$$

The result then follows from Theorem 5.3.1.

(ii) For $\epsilon > 0$, let $\{\hat{u}\}$ be $\frac{\epsilon}{2}$-optimal for $\{u\}$ at t , then from Definition 4.3.2,

$$(5.4.3) \qquad \{\hat{u}\} \in \mathcal{D} , \quad \{\hat{u}^{t-1}\} = \{u^{t-1}\}$$

and

$$(5.4.4) \qquad E_{\{\hat{u}\}}[L|z^t] \leq \hat{F}_t(z^t,\{u\}) + \frac{\epsilon}{2} \qquad \text{a.s. } P^t_{\{u\}} \quad .$$

Let

$$\{\hat{U}^{\epsilon}\} = \{\hat{U}^{Y,\mathcal{D}_0,\underline{\epsilon}}\}$$

where $\mathcal{D}_0$ is a selection class such that

$$(5.4.5) \qquad \{\hat{u}\} \in \mathcal{D}_0 \quad \text{and} \quad \frac{\epsilon}{2} = \sum_{\tau=t}^{T-1} \epsilon_{\tau} \quad .$$

Then from Theorem 5.3.1, (5.4.3), (5.4.5), and (5.4.4)

$$\{u^{t-1}\} \vee \{\hat{U}^{\epsilon}\} = \{\hat{u}^{t-1}\} \vee \{\hat{U}^{\epsilon}\} \in \mathcal{D}$$

and

$$E_{\{u^{t-1}\} \vee \{\hat{U}^{\epsilon}\}}[L|z^t] \leq E_{\{\hat{u}\}}[L|z^t] + \sum_{\tau=t}^{T-1} \epsilon_{\tau}$$

$$\leq \hat{F}_t(z^t,\{u\}) + \frac{\epsilon}{2} + \frac{\epsilon}{2} \qquad \text{a.s. } P^t_{\{u\}} \quad .$$

The result then follows from Definition 4.3.2.

iii) This is essentially a special case of (ii) where $t = 0$. Existence of ε-optimal controls at $t = 0$ is assured by Theorem 4.3.1 v).

Theorem 5.4.3. If the system has countably complete selction classes in Y, then the class of controls in Y satisfy the additional completeness conditions:

iv) For $t, \{u\} \in \mathcal{D}$, and $\varepsilon > 0$, there exists $\{\hat{U}^{\varepsilon}\}$ such that $\{u^{t-1}\} \vee \{\hat{U}^{\varepsilon}\}$ is ε-optimal for $\{u\}$ at t.

v) If there exists $\{\hat{u}\}$ optimal for $\{u\}$ at t, then there exists $\{\hat{U}\}$ such that $\{u^{t-1}\} \vee \{\hat{U}\}$ is optimal for $\{u\}$ at t.

vi) If there exists $\{\hat{u}\}$ optimal at t, then there exists $\{\hat{U}\}$ optimal at $t = 0$.

Proof: From Theorem 5.4.1 ii) $\mathcal{D}$ has the countable ε-lattice property. Thus from Theorem 4.3.4 for all $\{u\}$, t, and $\varepsilon > 0$ there exists $\{\hat{u}^{\varepsilon}\}$, ε-optimal for $\{u\}$ at t. The property iv) then follows from Theorem 5.4.2. ii). From Theorem 4.2.4 $\hat{F}_t(z^t, \{u\})$ is a submartingale for all $\{u\} \in \mathcal{D}$ and from Lemma 4.3.4

$$E_{\{\hat{u}\}}[L|z^t] = E[\hat{F}_{\tau}(z^{\tau}, \{\hat{u}\})|z^t] \qquad \text{a.s. } P^t_{\{\hat{u}\}}$$

for $\tau = t, t+1, \ldots, T$ where under the assumptions of v) $\{\hat{u}\}$ is optimal for $\{u\}$ at t. Let

$$\{\hat{U}\} = \{\hat{U}^{Y, \mathcal{D}_0, \underline{\varepsilon}}\}$$

where $\mathcal{D}_0$ is a selection class which contains $\{\hat{u}\}$ and $\underline{\varepsilon}$ is arbitrary. Then from Theorem 5.3.2, and (4.3.1) and (4.3.2) of Definition 4.3.1

$$\{u^{t-1}\} \vee \{\hat{U}\} = \{\hat{u}^{t-1}\} \vee \{\hat{U}\} \in \mathcal{D}$$

and

$$E_{\{u^{t-1}\} \vee \{\hat{U}\}}[L|z^t] = E_{\{\hat{u}\}}[L|z^t] = \hat{F}_t(z_t, \{u\}) \qquad \text{a.s. } P^t_{\{u\}} \quad .$$

The property v) then follows from Definition 4.3.1, and vi) is obtained from v) by letting $t = 0$.

Except for i) of Theorem 5.4.2, the results of the last two theorems are not constructive. They are interesting only in that they justify the restriction of attention to controls $\{U^Y\}$ in the sufficent statistic Y. The difficulty in construction lies in the fact that the required selection class is defined by its property of containing some optimal or ϵ-optimal $\{\hat{u}\}$, and that while the existence of $\{\hat{u}\}$ is assured either by assumption or by Theorem 4.3.4, there is no constructive method for obtaining $\{\hat{u}\}$.

Theorem 5.4.4. For a system with finitely complete selection classes in Y, let $\mathcal{D}_{\{u\}}$ be a selection class which contains $\{u\} \in \mathcal{D}$, and let

$$\hat{\ell}_t^Y(y_t,\{u\}) = (Y^{-1}P)^t_{\{u\}}\text{-ess}\inf_{\substack{\{\bar{u}\}\in\mathcal{D}\ \ni \\ \{\bar{u}^{t-1}\}=\{u^{t-1}\}}} \hat{\ell}_t^{Y,\mathcal{D}_{\{\bar{u}\}}}(y_t) \tag{5.4.6}$$

where $\hat{\ell}_t^{Y,\mathcal{D}_{\{u\}}}$ is defined by (5.2.30). Then $\hat{\ell}_t^Y$ satisfies

i) $\hat{\ell}_t^Y(y_t,\{u\})$ is $\mathcal{B}_{\mathcal{Y}_t}$-measurable for all $\{u\} \in \mathcal{D}$;

ii) for $\{\bar{u}\}\in\mathcal{D}$, $\{u\}\in\mathcal{D}$ such that $\{u^{t-1}\}=\{\bar{u}^{t-1}\}$

$$\hat{\ell}_t(y_t,\{u\}) = \hat{\ell}_t(y_t,\{\bar{u}\}) \qquad \text{a.s. } (Y^{-1}P)^t_{\{u\}} \quad ;$$

and

iii) for all $\{u\} \in \mathcal{D}$

$$\hat{F}_t(z^t,\{u\}) = \hat{\ell}_t^Y(Y_t(z^t,u^{t-1}(z^{t-1}));\{u\}) + E_{\{u\}}[\sum_{\tau=1}^t c_\tau^V|z^t] \quad \text{a.s. } P^t_{\{u\}} . \tag{5.4.7}$$

Proof: From Lemma 5.2.4, the functions $\hat{\ell}^{Y,\mathcal{D}_{\{\bar{u}\}}}(y_t)$ are non-negative and $\mathcal{B}_{\mathcal{Y}_t}$-measurable. Thus the P-ess inf in (5.4.6) is well-defined. The property i) follows from i) of Definition A.2.1, ii) is obvious from the definition (5.4.6). Since $\{\bar{u}\} \in \mathcal{D}_{\{\bar{u}\}}$ from (5.2.21) of Definition 5.2.3

$$\{\bar{u}^{t-1}\} \vee \{\hat{U}^{\epsilon^n}\} \in \mathcal{D} .$$

Thus from (5.3.16) of Lemma 5.3.4

$$\hat{F}_t(z^t,\{\bar{u}\}) \leq \hat{\ell}^{Y,\mathcal{D}_{\{\bar{u}\}}}(Y_t(z^t,\bar{u}^{t-1}(z^{t-1}))) + E_{\{\bar{u}\}}[\sum_{\tau=1}^t c_\tau^V|z^t] \quad \text{a.s. } P^t_{\{\bar{u}\}} .$$

Thus for all $\{\bar{u}\} \in \mathcal{D}$ such that $\{\bar{u}^{t-1}\} = \{u^{t-1}\}$

$$\hat{F}_t(z^t,\{u\}) = \hat{F}_t(z^t,\{\bar{u}\}) \leq \hat{\ell}^{Y,\mathcal{D}}{}^{\{\bar{u}\}}(Y_t(z^t,u^{t-1}(z^{t-1}))) + E_{\{u\}}[\sum_{\tau=1}^{t} c_\tau^V | z^t] \quad \text{a.s. } P_{\{u\}}^t .$$

It follows from iii) of Definition A.2.1 that

$$\hat{F}_t(z^t,\{u\}) \leq P_{\{u\}}^t\text{-ess inf}_{\substack{\bar{u} \in \mathcal{D} \ni \\ \{\bar{u}^{t-1}\} = \{u^{t-1}\}}} \{\hat{\ell}^{Y,\mathcal{D}}{}^{\{\bar{u}\}}(Y_t(z^t,u^{t-1}(z^{t-1}))) + E_{\{u\}}[\sum_{\tau=1}^{t} c_\tau^V | z^t]\} \quad \text{a.s. } P_{\{u\}}^t . \tag{5.4.8}$$

From (5.3.17) of Lemma 5.3.4

$$\hat{\ell}^{Y,\mathcal{D}}{}^{\{\bar{u}\}}(Y_t(z^t,\bar{u}^{t-1}(z^{t-1}))) + E_{\{\bar{u}\}}[\sum_{\tau=1}^{t} c_\tau^V | z^t] \leq E_{\{\bar{u}\}}[L|z^t] \quad \text{a.s. } P_{\{\bar{u}\}}^t .$$

Thus from Lemma A.2.8 and Definition 4.2.3

$$\begin{aligned} &P_{\{u\}}^t\text{-ess inf}_{\substack{\{\bar{u}\} \in \mathcal{D} \ni \\ \{\bar{u}^{t-1}\} = \{u^{t-1}\}}} \{\hat{\ell}_t^{Y,\mathcal{D}}{}^{\{\bar{u}\}}(Y_t(z^t,u^{t-1}(z^{t-1}))) + E_{\{u\}}[\sum_{\tau=1}^{t} c_\tau^V | z^t]\} \\ &\leq P_{\{u\}}^t\text{-ess inf}_{\substack{\{\bar{u}\} \in \mathcal{D} \ni \\ \{\bar{u}^{t-1}\} = \{u^{t-1}\}}} E_{\{\bar{u}\}}[L|z^t] = \hat{F}_t(z^t,\{u\}) \quad \text{a.s. } P_{\{u\}}^t . \end{aligned} \tag{5.4.9}$$

Thus from (5.4.8), (5.4.9), and Lemma A.2.11

$$\begin{aligned} \hat{F}_t(z^t,\{u\}) &= P_{\{u\}}^t\text{-ess inf}_{\substack{\{\bar{u}\} \in \mathcal{D} \ni \\ \{\bar{u}^{t-1}\} = \{u^{t-1}\}}} \{\hat{\ell}_t^{Y,\mathcal{D}}{}^{\{\bar{u}\}}(Y_t(z^t,u^{t-1}(z^{t-1}))) \\ &\quad + E_{\{u\}}[\sum_{\tau=1}^{t} c_\tau^V | z^t]\} = E_{\{u\}}[\sum_{\tau=1}^{t} c_\tau^V | z^t] + P_{\{u\}}^t\text{-ess inf}_{\substack{\{\bar{u}\} \in \mathcal{D} \ni \\ \{\bar{u}^{t-1}\} = \{u^{t-1}\}}} \hat{\ell}_t^{Y,\mathcal{D}}{}^{\{\bar{u}\}}(Y_t(z^t,u^{t-1}(z^{t-1}))) \\ &\qquad \text{a.s. } P_{\{u\}}^t . \end{aligned} \tag{5.4.10}$$

From (5.4.6) and Lemma A.2.13

$$(5.4.11)\quad P^t_{\{u\}}\text{-ess inf}_{\substack{\{\bar u\}\in \mathcal{D} \ni \\ \{\bar u^{t-1}\} = \{u^{t-1}\}}} \hat{\ell}_t^{Y,\mathcal{D}\{\bar u\}}(Y_t(z^t,u^{t-1}(z^{t-1}))) = \hat{\ell}_t^{Y}(Y_t(z^t,u^{t-1}(z^{t-1}))) \text{ a.s.} P^t_{\{u\}}.$$

(5.4.7) then follows from (5.4.10) and (5.4.11).

Theorem 5.4.5. Let the system have countably complete selection classes in Y, let $\{\hat u\} \in \mathcal{D}$, $\{u\} \in \mathcal{D}$ satisfy

$$(5.4.12)\qquad \{\hat u^{t-1}\} = \{u^{t-1}\},$$

and for $\{\bar u\} \in \mathcal{D}$ such that

$$(5.4.13)\qquad \{\hat u^{t-1}\} = \{\bar u^{t-1}\}$$

let $\mathcal{D}_{\{\hat u\},\{\bar u\}}$ be a selection class that contains $\{\hat u\}$ and $\{\bar u\}$. Then $\{\hat u\}$ is optimal for $\{u\}$ at t if and only if for $\tau = t,t+1,\ldots,T-1$ and all $\{\bar u\}$ which satisfies (5.4.13)

$$(5.4.14)\qquad \begin{aligned} &E_{\hat u}[\sum_{\alpha=1}^{\tau} c^V_\alpha + \hat L_\tau^{Y,\mathcal{D}\{\hat u\},\{\bar u\}}(\hat Y_\tau,\hat u_\tau(z^\tau))\mid z^t] \\ &= E_{\hat u}[\sum_{\alpha=1}^{\tau} c^V_\alpha + \hat{\ell}_\tau^{Y,\mathcal{D}\{\hat u\},\{\bar u\}}(\hat Y_\tau)\mid z^t] \qquad \text{a.s. } P^t_{\{\hat u\}}. \end{aligned}$$

Proof: Let $\{\hat u\}$ satisfy (5.4.14), then from Lemma 5.3.6

$$E_{\{\hat u\}}[L\mid z^t] \le E_{\{\bar u\}}[L\mid z^t] \qquad \text{a.s. } P^t_{\{\hat u\}}$$

for all $\{\bar u\}$ which satisfy (4.5.13). Thus from the definition of $\hat F_t(z^t,\{\hat u\})$

$$E_{\hat u}[L\mid z^t] \le \hat F_t(z^t,\{\hat u\}) \le E_{\{\hat u\}}[L\mid z^t] \qquad \text{a.s. } P^t_{\{u\}}.$$

It follows then that (4.3.2) holds, and hence $\{\hat u\}$ is optimal for $\{u\}$ at t.

Suppose then that $\{\hat u\}$ is optimal for $\{u\}$ at t. From Theorem 5.4.1 ii) the system has the countable ϵ-lattice property. Thus from Theorem 4.2.4 it follows that $\hat F_t(z^t,\{\hat u\})$ is a submartingale. From Lemma

4.3.4 $\{\hat{u}\}$ satisfies

$$E_{\{\hat{u}\}}[L|z^t] = \hat{F}_t(z^t,\{\hat{u}\}) = E_{\{\hat{u}\}}[\hat{F}_\tau(z^\tau,\{\hat{u}\})|z^t] \qquad \text{a.s. } P^t_{\{\hat{u}\}}$$

for $\tau = t,t+1,\ldots,T$. It follows then from Lemma 5.3.3 that

$$E_{\{\hat{u}\}}[\sum_{\alpha=1}^{\tau} c^V_\alpha + L_\tau^{Y,\mathcal{D}\{\hat{u}\},\{\bar{u}\},\varepsilon^n}(\hat{Y}_\tau,\hat{u}_\tau)|z^t]$$

$$= E_{\{\hat{u}\}}[\sum_{\alpha=1}^{\tau} c^V_\alpha + \ell_\tau^{Y,\mathcal{D}\{\hat{u}\},\{\bar{u}\},\varepsilon^n}(\hat{Y}_\tau)|z^t] \qquad \text{a.s. } P^t_{\{\hat{u}\}}$$

for $\tau = t,t+1,\ldots,T$. The result (5.4.14) is obtained from (5.2.30) and (5.2.31) of Lemma 5.2.4 and the monotone convergence theorem for conditional expectations.

Theorem 5.4.6. Let the system have finitely complete selection classes, let $\{\hat{u}\}$, $\{u\}$, and $\mathcal{D}_{\hat{u},\{\bar{u}\}}$ be as in Theorem 5.4.5, and assume that

(5.4.15) $$E_{\{\hat{u}\}}[L] < \infty .$$

Then $\{\hat{u}\}$ is optimal for $\{u\}$ at t if and only if

(5.4.16)
$$\hat{L}_\tau^{Y,\mathcal{D}\{\hat{u}\},\{\bar{u}\}}(Y(z^\tau,\hat{u}^{\tau-1}(z^{\tau-1})) , \hat{u}_\tau(z^\tau))$$
$$= \hat{\ell}_\tau^{Y,\mathcal{D}\{\hat{u}\},\{\bar{u}\}}(Y_\tau(z^\tau,\hat{u}^{\tau-1}(z^{\tau-1}))) \qquad \text{a.s. } P^\tau_{\{\hat{u}\}}$$

for $\tau = t,t+1,\ldots,T-1$ and all $\{\bar{u}\}$ which satisfies (5.4.13) .

Proof: Clearly the condition (5.4.16) implies that (5.4.14) of the previous theorem holds. Thus by the argument of the previous proof, (5.4.16) implies that $\{\hat{u}\}$ is optimal for $\{u\}$ at t .

Assume then that $\{\hat{u}\}$ is optimal for $\{u\}$ at t . From Theorem 5.4.1 i) $\mathcal{D}$ has the finite ε-lattice property, and then from Theorem 4.2.3 and (5.4.15) $\hat{F}_t(z^t,\{\hat{u}\})$ is a submartingale. Thus from (5.4.15), Lemma 4.3.5 and Definition 4.3.1

$$E_{\{\hat{u}\}}[L|z^s] = \hat{F}_s(z^s,\{\hat{u}\}) = E_{\{\hat{u}\}}[\hat{F}_\tau(z^\tau,\{\hat{u}\}) |z^s] \qquad \text{a.s. } P^s_{\{\hat{u}\}}$$

for $t \leq s \leq \tau \leq T$. As before Lemma 5.3.3 holds. From (5.3.10)

$$E_{\{\hat{u}\}}[\sum_{\alpha=1}^{\tau} c_\alpha^V | z^\tau] + L_\tau^{Y,\mathcal{D}_{\{\hat{u}\},\{\bar{u}\}},\underline{\varepsilon}^n}(\hat{Y}_\tau, \hat{u}_\tau(z^\tau))$$

$$= E_{\{\hat{u}\}}[\sum_{\alpha=1}^{\tau} c_\alpha^V | z^\tau] + \ell_\tau^{Y,\mathcal{D}_{\{\hat{u}\},\{\bar{u}\}},\underline{\varepsilon}^n}(\hat{Y}_\tau) \qquad \text{a.s. } P^\tau_{\{\hat{u}\}} .$$

From (5.4.15) the first term on each side is a.s. finite. The result (5.4.16) then follows from (5.2.30) and (5.2.31) of Lemma 5.2.4.

Theorem 5.4.7. Let the system have finitely complete selection classes, let $\{\hat{u}\}$, $\{u\}$, and $\mathcal{D}_{\{\hat{u}\},\{\bar{u}\}}$ be as in Theorem 5.4.5, and assume that $\{\hat{u}\}$ satisfies

$$(5.4.17) \qquad L_s^{Y,\mathcal{D}_{\{\hat{u}\},\{\bar{u}\}},\underline{\varepsilon}}(\hat{Y}_s, \hat{u}_s(z^s)) \leq \ell_s^{Y,\mathcal{D}_{\{\hat{u}\},\{\bar{u}\}},\underline{\varepsilon}}(\hat{Y}_s) \qquad \text{a.s. } P^s_{\{\hat{u}\}}$$

for $s = t, t+1, \ldots, T-1$ and all $\{\bar{u}\}$ which satisfies (5.4.13) . Then $\{\hat{u}\}$ is ε-optimal for $\{u\}$ at t where

$$(5.4.18) \qquad \varepsilon = \sum_{\tau=t}^{T-1} \varepsilon_\tau .$$

Proof: From Lemma 5.3.5 with $\mathcal{D}_0 = \mathcal{D}_{\{\hat{u}\},\{\bar{u}\}}$

$$E_{\{\hat{u}\}}[L|z^t] \leq E_{\{\bar{u}\}}[L|z^t] + \sum_{\tau=t}^{T-1} \varepsilon_\tau \qquad \text{a.s. } P^t_{\{\hat{u}\}} .$$

Since this holds for all $\{\bar{u}\}$ which satisfy (5.4.13) , it follows from (5.4.18) and the definition of $\hat{F}_t(z^t, \{\hat{u}\})$ that

$$E_{\hat{u}}[L|z^t] - \varepsilon \leq \hat{F}_t(z^t, \{\hat{u}\}) \qquad \text{a.s. } P^t_{\{\hat{u}\}} .$$

The result then follows from Definition 4.3.2.

Lemma 5.4.1. Let $\mathcal{Y}$ and $\mathcal{U}$ be complete separable metric spaces, let $\mathcal{B}_\mathcal{Y}$ and $\mathcal{B}_\mathcal{U}$ be the Borel σ-fields over the topologies of the spaces, let $L(y,u)$ be non-negative (possibly $+\infty$) and $\mathcal{B}_\mathcal{Y} \times \mathcal{B}_\mathcal{U}$-measureable, let $\bar{\rho}$ be a σ-finite measure on $(\mathcal{Y}, \mathcal{B}_\mathcal{Y})$ and let $\varepsilon > 0$. Then there exists $U^\varepsilon(y)$ and $\tilde{\ell}(y)$ $\mathcal{B}_\mathcal{Y}$-measurable which satisfy

$$(5.4.19) \qquad \tilde{\ell}(y) = \inf_{u \in \mathcal{U}} L(y,u) \qquad \text{a.e. } \bar{\rho}$$

$$(5.4.20) \qquad \chi_B(y)L(y,U^\epsilon(y)) = \chi_B(y)\tilde{\ell}(y) \qquad \text{a.e. } \bar{\rho}$$

$$(5.4.21) \qquad \chi_{B^c}(y)L(y,U^\epsilon(y)) \le \chi_{B^c}(y)\tilde{\ell}(y) + \epsilon \qquad \text{a.s. } \bar{\rho}$$

where $B \in \mathcal{B}_{\mathcal{Y}}$ and

$$(5.4.22) \qquad B = \{y \mid \exists u^* \in \mathcal{U} \ni L(y,u^*) = \tilde{\ell}(y)\} \qquad \text{a.e. } \bar{\rho} \ .$$

Proof: This follows from a selection theorem of Brown and Purves [3] . Theorem 2 of [3] can easily be extended to real-valued functions which admit the value $+\infty$. It follows then that there exist $\bar{U}^\epsilon(y), \bar{\ell}(y)$, and $\bar{B}$ absolutely measurable which satisfy (5.4.19)-(5.4.22) for all y . From the definition of absolute measurability, $\bar{U}^\epsilon(y)$ and $\bar{\ell}(y)$ are $\bar{\mathcal{B}}_{\mathcal{Y}}$-measurable and $\bar{B} \in \bar{\mathcal{B}}_{\mathcal{Y}}$ where $\bar{\mathcal{B}}_{\mathcal{Y}}$ is the completion of $\mathcal{B}_{\mathcal{Y}}$ with respect to $\bar{\rho}$. It follows then that there exists $U^\epsilon(y)$ and $\tilde{\ell}(y)$ $\mathcal{B}_{\mathcal{Y}}$-measurable and $B \in \mathcal{B}_{\mathcal{Y}}$ which satisfy

$$(5.4.23) \qquad \begin{aligned} \bar{\ell}(y) &= \tilde{\ell}(y) && \text{a.e. } \bar{\rho} \\ \bar{U}^\epsilon(y) &= U(y) && \text{a.e. } \bar{\rho} \\ \bar{B} &= B && \text{a.e. } \bar{\rho} \ . \end{aligned}$$

The equations (5.4.19)-(5.4.22) hold except for those values of y on which (5.4.23) fail to hold.

<u>Definition</u> 5.4.3. The class $\mathcal{D}$ of control laws is <u>universal with respect to the statistic</u> Y provided

$$(5.4.24) \qquad \{u^{t-1}\} \vee \{U\} \in \mathcal{D}$$

for all $\{u\} \in \mathcal{D}$, t , and control laws $\{U\}$ in Y .

<u>Theorem</u> 5.4.8 . Let Y be a statistic sufficient for the control system (Definitions 3.2.2 and 3.2.6), let $\mathcal{Y}_{t+1}$ and $\mathcal{U}_t$ be complete separable metric spaces for $t = 0,1,2,\ldots,T-1$, and let $\mathcal{D}$ be universal in Y . Then for

all ρ , the class $\mathcal{D}_\rho$ (Definition 3.2.10) is a selection class (Definition 5.2.3). Further, for

$$\underline{\varepsilon} = (\varepsilon_0, \varepsilon_1, \ldots, \varepsilon_{T-1}) \tag{5.4.25}$$

the control law $\{\hat{U}^{Y,\mathcal{D}_\rho,\underline{\varepsilon}}\}$ and the measurable functions $L_t^{Y,\mathcal{D}_\rho,\underline{\varepsilon}}(Y_t,u_t)$, $\tilde{\ell}_t^{Y,\mathcal{D}_\rho,\underline{\varepsilon}}(y_t)$, and $\ell_t^{Y,\mathcal{D}_\rho,\underline{\varepsilon}}(y_t)$ can be constructed by backward induction from the equations

$$L_T = \ell_T = 0 \ , \tag{5.4.26}$$

$$c_{t+1}^Y(y_{t+1},u_t) = \int_{\mathcal{V}_{t+1}} c_{t+1}^V(v_{t+1},u_t)G_{t+1}^{Y,V}(y_{t+1},dv_{t+1}) \tag{5.4.27}$$

$$L_t(y_t,u_t) = \int_{\mathcal{Y}_{t+1}} [\ell_{t+1}(y_{t+1}) + c_{t+1}^Y(y_{t+1},u_t)]K_t^Y(y_t,u_t;dy_{t+1}) \tag{5.4.28}$$

$$\tilde{\ell}_t(y_t) = \inf_{u_t \in \mathcal{U}_t} L_t(y_t,u_t) \qquad \text{a.e. } \rho_t \tag{5.4.29}$$

$$B_t = \{y_t | \exists u_t^* \in \mathcal{U}_t \ni L_t(y_t,u_t^*) = \tilde{\ell}_t(y_t)\} \qquad \text{a.e. } \rho_t \tag{5.4.30}$$

$$B_t \in \mathcal{B}_{\mathcal{Y}_t} \tag{5.4.31}$$

$$L_t(y_t,\hat{U}_t(y_t)) = \tilde{\ell}_t(y_t) \quad \text{on } B_t \qquad \text{a.e. } \rho_t \tag{5.4.32}$$

$$L_t(y_t,\hat{U}_t(y_t)) \le \tilde{\ell}_t(y_t) + \varepsilon_t \quad \text{on } B_t^c \qquad \text{a.e. } \rho_t \tag{5.4.33}$$

$$\ell_t(y_t) = L_t(y_t,\hat{U}_t(y_t)) \ . \tag{5.4.34}$$

Proof: From Lemma 5.4.1 it follows easily by backward induction that there exist functions which satisfy (5.4.26)-(5.4.34). The measurability of c_{t+1}^Y and $L_t(y_t,u_t)$ defined by (5.4.27) and (5.4.28) follows from Lemma 5.2.1. The exceptimal sets of (5.4.29), (5.4.30), (5.4.32), and (5.4.33) have ρ_t-measure zero. Thus from Definition 3.2.10 such sets have $(Y^{-1}P)_{\{u\}}^t$ measure zero for all $\{u\} \in \mathcal{D}_\rho$. It follows then that the functions L_t , $\tilde{\ell}_t$, ℓ_t and $\hat{U}_t$ satisfy (5.2.23)-(5.2.28) for all $\{u\} \in \mathcal{D}_\rho$.

It remains to be shown that the $\hat{U}_t^{\varepsilon_t^n}(y_t)$ can be selected in such a way that $\ell_t^{Y,\mathcal{D}_\rho,\underline{\varepsilon}^n}$ and $L^{Y,\mathcal{D}_\rho,\underline{\varepsilon}^n}$ are non-increasing in n . The proof will be by forward induction on n . Denoting the functions by ℓ_t^n and L_t^n , for n fixed assume that

$$L_t^{m+1}(y_t,u_t) \leq L_t^m(y_t,u_t) \tag{5.4.35}$$

$$\ell_t^{m+1}(y_t) \leq \ell_t^m(y_t) \tag{5.4.36}$$

for all t , y_t , u_t and $m=1,2,\ldots,n-1$. It will be shown by backward induction on t that (5.4.35) and (5.4.36) hold for $m = n$. Since $L_T^n = \ell_T^n = 0$ for all n , (5.4.35) and (5.4.36) are trivial for $t = T$. Assume then that

$$L_{t+1}^{n+1}(y_{t+1},u_{t+1}) \leq L_{t+1}^n(y_{t+1},u_{t+1})$$

$$\ell_{t+1}^{n+1}(y_{t+1}) \leq \ell_{t+1}^n(y_{t+1}) \tag{5.4.37}$$

for t fixed and all y_{t+1} , u_{t+1} . From (5.4.37) and (5.4.28)

$$L_t^{n+1}(y_t,u_t) \leq L_t^n(y_t,u_t) \tag{5.4.38}$$

for all y_t and u_t . From Lemma 5.4.1 let $\tilde{U}_t^{n+1}(y_t)$ satisfy

$$\chi_{B_t}(y_t)L_t^{n+1}(y_t,\tilde{U}_t^{n+1}(y_t)) = \chi_{B_t}(y_t) \inf_{u_t \in \mathcal{U}_t} L_t^{n+1}(y_t,u_t) \qquad \text{a.e. } \rho_t \tag{5.4.39}$$

$$\chi_{B_t^c}(y_t)L_t^{n+1}(y_{t+1},\tilde{U}_t^{n+1}(y_t)) \leq \chi_{B^c}(y_t) \inf_{u_t \in \mathcal{U}_t} L_t^{n+1}(y_t,u_t) + \frac{1}{n+1} \qquad \text{a.e. } \rho_t \tag{5.4.40}$$

Define $\hat{U}_t^{n+1}$ as follows

$$\hat{U}_t^{n+1}(y_t) = \tilde{U}_t^{n+1}(y_t) \quad \text{if} \quad L_t^{n+1}(y_t,\tilde{U}_t^{n+1}(y_t)) \leq L_t^n(y_t,\hat{U}_t^n(y_t)) \tag{5.4.41}$$

$$\hat{U}_t^{n+1}(y_t) = \hat{U}_t^n(y_t) \quad \text{if} \quad L_t^{n+1}(y_t,\tilde{U}_t^{n+1}(y_t)) > L_t^n(y_t,\hat{U}_t^n(y_t)) \tag{5.4.42}$$

The function $\ell_t^{n+1}(y_t)$ then is defined by (5.4.34). For y_t as in

(5.4.41)

$$\ell_t^{n+1}(y_t) = L_t^{n+1}(y_t, \hat{U}_t^{n+1}(y_t)) = L_t^{n+1}(y_t, \tilde{U}_t^{n+1}(y_t))$$

$$\leq L_t^n(y_t, \hat{U}_t^n(y_t)) = \ell_t^n(y_t) \ .$$

For y_t as in (5.4.42), from (5.4.38)

$$\ell_t^{n+1}(y_t) = L_t^{n+1}(y_t, \hat{U}_t^{n+1}(y_t)) = L_t^{n+1}(y_t, \hat{U}_t^n(y_t))$$

$$\leq L_t^n(y_t, \hat{U}_t^n(y_t)) = \ell_t^n(y_t) < L_t^{n+1}(y_t, \tilde{U}_t^{n+1}(y_t)) \ .$$

Thus for all y_t

$$\ell_t^{n+1}(y_t) \leq \ell_t^n(y_t) \ ,$$

$$L_t^{n+1}(y_t, \hat{U}_t^{t+1}(y_t)) \leq L_t^{n+1}(y_t, \tilde{U}_t(y_t)) \ ,$$

and clearly (5.4.39) and (5.4.40) are also satisfies by $\hat{U}_t^{n+1}(y_t)$. Since the functions involved are measurable, $\hat{U}_t^{n+1}(y_t)$ defined by (5.4.41), (5.4.42) is measurable.

Corollary 5.4.1. Let Y be a statistic sufficient for the control system, let $\mathcal{Y}_{t+1}$, $\mathcal{U}_t$, $t = 0,1,\ldots,T-1$ be complete separable metric spaces, and let $\mathcal{D}$ be universal in Y . Then the system has countably complete selection classes in Y .

Proof: Let $\bar{\mathcal{D}} = \{\{\bar{u}\}_1, \{\bar{u}\}_2, \ldots\}$ be a countable class of laws. For $B \in \mathcal{B}_{\mathcal{Y}_t}$, let

$$\rho_t(B) = \sum_{n=1}^{\infty} \frac{1}{2^n} (Y^{-1}P)^t_{\{\bar{u}\}_n}(B) \ .$$

Then ρ_t is a finite measure on $(\mathcal{Y}_t, \mathcal{B}_{\mathcal{Y}_t})$ and

$$\bar{\mathcal{D}} \subset \mathcal{D}_\rho \ .$$

From Theorem 5.4.8 , $\mathcal{D}_\rho$ is a selection class and the result follows from the Definition 5.4.2.

Corollary 4.5.2. Let $\mathcal{X}_0, \mathcal{X}_t, Z_t, t=1,\ldots,T$ be finite dimensional vector spaces, let $\mathcal{U}_t, t=0.1,\ldots,T-1$ be complete separable metric spaces, let L have the form (1.2.32), and let all control laws be admissible. Then the system has countably complete selection classes for $Y = I$, the conditional loss function $\hat{F}_t$ is characterized by the properties i) and ii) of Theorem 4.2.1, necessary and sufficient conditions for optimality of a law $\{\hat{u}\}$ are given by Theorems 4.3.8, 5.4.5 and 5.4.6 and necessary conditions for ϵ-optimality are given by Theorem 5.4.7.

Proof: From Theorem 2.2.3 there exist stochastic kernels G_t and K_t which satisfy (2.1.1) and (2.1.2) for all $\{u\}$. Thus from Lemma 3.2.4 the identity statistic is sufficient for the identity structure. From Lemma 3.2.3 a loss function of the form (1.2.32) has identity structure. Thus I is sufficient for the control system (Definition 3.2.6). The range spaces of I_t, $\mathcal{Y}_t = Z^t \times \mathcal{U}^{t-1}$ are complete, separable, metric spaces. Thus from Corollary 5.4.1 the system has countably complete selection classes and from Theorem 5.4.1 ii) the system has the countable ϵ-lattice property. The characterization of $\hat{F}_t$ follows from Theorem 4.2.5 and the remaining results follow from the cited theorems.

5.5 Single selection class

This section contains the only result, Theorem 5.5.5, which is truly applicable in practical situations. In addition to optimality properties obtained in the previous section, it deals with the stronger universal optimality of Definition 5.5.2.

Definition 5.5.1. The system is said to have a single selection class in the sufficient statistic Y provided $\mathcal{D}$ is a selection class for Y.

In the case of a single selection class the superscript $\mathcal{D}$ on the functions $\ell_t^{Y,\mathcal{D},\underline{\epsilon}}$, $L_t^{Y,\mathcal{D},\underline{\epsilon}}$, $\hat{\ell}_t^{Y,\mathcal{D}}$, and $\ell_t^{Y,\mathcal{D}}$ is unnecessary and will usually be omitted.

Definition 5.5.2. A control law $\{\hat{U}^Y\}$ in Y is a universally optimal law provided for all $\{u\} \in \mathcal{D}$ and all t, $\{u^{t-1}\} \vee \{\hat{U}^Y\}$ is optimal for $\{u\}$ at t.

Definition 5.5.3. A control law $\{\hat{U}^{Y,\epsilon}\}$ in Y is a universally ϵ-optimal law provided for all $\{u\} \in \mathcal{D}$ and all t, $\{u^{t-1}\} \vee \{\hat{U}^{Y,\epsilon}\}$ is ϵ-optimal for $\{u\}$ at t. Further, if there exists $\{\hat{u}\}$ optimal for $\{u\}$ at t, then $\{u^{t-1}\} \vee \{\hat{U}^{Y,\epsilon}\}$ is optimal for $\{u\}$ at t.

First an obvious property of the single selection class system is stated.

Lemma 5.5.1. If the system has a single selection class in Y, then it has countably complete selection classes and the countable ϵ-lattice property. Further, $\hat{F}_t(z^t,\{u\})$ is a submartingale for all $\{u\} \in \mathcal{D}$.

Proof: The results follow from Definition 5.4.2, Definition 5.5.1, Theorem 5.4.1 ii) and Theorem 4.2.4.

Theorem 5.5.1. Let the sufficient statistic Y be dominated by ρ (Definition 3.2.11), let $\mathcal{Y}_{t+1}$ and $\mathcal{U}_t$ for $t = 0,1,\ldots,T-1$ be complete separable metric spaces, and let $\mathcal{D}$ be universal in Y. Then the system has a single selection class, and the control laws $\{\hat{U}^{Y,\underline{\epsilon}}\}$ can be constructed by backward induction for the equations (5.4.26)-(5.4.34).

Proof: From Definition 3.2.11

$$\mathcal{D}_\rho = \mathcal{D} .$$

The result follows from Theorem 5.4.8.

Theorem 5.5.2. Let the system have a single selection class for Y, and let

$$\{\hat{U}^{Y,\epsilon}\} = \{\hat{U}^{Y,\mathcal{D},\underline{\epsilon}}\} \tag{5.5.1}$$

where

$$\epsilon = \sum_{\tau=0}^{T-1} \epsilon_\tau . \tag{5.5.2}$$

Then the control laws $\{\hat{U}^{Y,\epsilon}\}$ are universally ϵ-optimal (Definition 5.5.3).

Proof: Since $\mathcal{D} = \mathcal{D}_0$ is a selection class, from Theorem 5.3.1, (5.5.1), and (5.5.2)

$$E_{\{u^{t-1}\} \vee \{\hat{U}^{Y,\epsilon}\}}[L|z^t] \leq E_{\{u\}}[L|z^t] + \epsilon \qquad \text{a.s. } P^t_{\{u\}}$$

for all $\{u\} \in \mathcal{D}$ and $0 \leq t \leq T$. It follows then from property iii) of the definition of the P-ess inf (Definition A.2.1) and Definition 4.2.3 that

$$E_{\{u^{t-1}\} \vee \{\hat{U}^{Y,\epsilon}\}}[L|z^t] - \epsilon \leq P^t_{\{u\}}\text{-ess inf}_{\substack{\{\bar{u}\} \in \mathcal{D} \ni \\ \{\bar{u}^{t-1}\} = \{u^{t-1}\}}} E_{\{\bar{u}\}}[L|z^t]$$

$$= \hat{F}_t(z^t, \{u\}) \qquad \text{a.s. } P^t_{\{u\}} \quad .$$

Thus from Definition 4.3.2, $\{u^{t-1}\} \vee \{\hat{U}^{Y,\epsilon}\}$ is ϵ-optimal for $\{u\}$ at t. This holds for all $\{u\} \in \mathcal{D}$ and t. Suppose then that for $\{u\}$ and t fixed, there exists $\{\hat{u}\}$ optimal for $\{u\}$ at t. From Lemma 5.5.1 $\hat{F}_t(z^t, \{\hat{u}\})$ is a submartingale, and hence from Lemma 4.3.4

$$E_{\{\hat{u}\}}[L|z^t] = E[\hat{F}_\tau(z^\tau, \{\hat{u}\})|z^t] \qquad \text{a.s. } P^t_{\{\hat{u}\}}$$

for $\tau = t, t+1, \ldots, T$. Then from Theorem 5.3.2

$$E_{\{u^{t-1}\} \vee \{\hat{U}^{Y,\epsilon}\}}[L|z^t] = E_{\{\hat{u}\}}[L|z^t] = \hat{F}_t(z^t, \{\hat{u}\}) \qquad \text{a.s. } P^t_{\{\hat{u}\}}$$

and it follows that $\{u^{t-1}\} \vee \{\hat{U}^{Y,\epsilon}\}$ is optimal for $\{u\}$ at t (Definition 4.3.1). It has now been established that $\{\hat{U}^{Y,\epsilon}\}$ is universally ϵ-optimal.

Theorem 5.5.3. If the system has a single selection class for Y, then the functions

$$(5.5.3) \qquad L^{Y,\epsilon}_t = L^{Y,\mathcal{D},\underline{\epsilon}}_t, \; \ell^{Y,\epsilon}_t = \ell^{Y,\mathcal{D},\underline{\epsilon}}_t, \; \hat{L}^Y_t = \hat{L}^{Y,\mathcal{D}}_t, \; \hat{\ell}^Y_t = \hat{\ell}^{Y,\mathcal{D}}_t$$

where

$$(5.5.4) \qquad \epsilon = \sum_{\tau=0}^{T-1}$$

satisfy

i) for all $\{u\} \in \mathcal{D}$

$$(5.5.5) \quad \hat{F}_t(z^t,\{u\}) = \hat{\ell}^Y_t(Y_t(z^t,u^{t-1}(z^{t-1}))) + E_{\{u\}}[\sum_{\tau=1}^{t} c^V_\tau | z^t] \qquad \text{a.s. } P^t_{\{u\}} \;;$$

ii) $\{\hat{u}\}$ is optimal for $\{u\}$ at t if and only if

$$(5.5.6) \quad \{\hat{u}^{t-1}\} = \{u^{t-1}\}$$

and

$$(5.5.7) \quad \begin{aligned} &E_{\{\hat{u}\}}[\sum_{\alpha=1}^{T} c^V_\alpha + \hat{L}^Y_\tau(\hat{Y}_\tau,\hat{u}_\tau(z^\tau))|z^t] \\ &= E_{\{\hat{u}\}}[\sum_{\tau=1}^{T} c^V_\alpha + \hat{\ell}^Y_\tau(\hat{Y}_\tau)|z^t] \qquad \text{a.s. } P^t_{\{\hat{u}\}} \end{aligned}$$

for $\tau = t,t+1,\ldots,T-1$;

iii) if $\{\hat{u}\}$ satisfies (5.5.6) and

$$(5.5.8) \quad E_{\{\hat{u}\}}[L] < \infty \;,$$

then $\{\hat{u}\}$ is optimal for $\{u\}$ at t if and only if

$$(5.5.9) \quad \hat{L}^Y_\tau(\hat{Y}_\tau,\hat{u}_\tau(z^t)) = \hat{\ell}^Y_\tau(\hat{Y}_\tau) \qquad \text{a.s. } P^\tau_{\{\hat{u}\}}$$

for $\tau = t,t+1,\ldots,T-1$; and

iv) if $\{\hat{u}^\epsilon\}$ satisfies (5.5.6) and

$$(5.5.10) \quad L^{Y,\epsilon}_\tau(\hat{Y}_\tau,\hat{u}^\epsilon_\tau(z^t)) \leq \ell^{Y,\epsilon}_\tau(\hat{Y}_\tau) \qquad \text{a.s. } P^\tau_{\{\hat{u}^\epsilon\}}$$

for $\tau = t,t+1,\ldots,T-1$, then $\{\hat{u}^\epsilon\}$ is ϵ-optimal for $\{u\}$ at t .

Proof: For all $\{\bar{u}\} \in \mathcal{D}$, let

$$(5.5.11) \quad \mathcal{D}_{\{\bar{u}\}} = \mathcal{D} \;.$$

Then from the definition (5.4.6) of $\hat{\ell}^Y_t(y_t,\{u\})$ and Lemma A.2.7

$$\hat{\ell}_t(y_t,\{u\}) = \hat{\ell}_t(y_t)$$

for all $\{u\} \in \mathcal{D}$. Property i) then follows from (5.4.7) of Theorem 5.4.4.

The properties ii-iv) follow from Theorems 5.4.5-5.4.7 where

(5.5.12) $$\mathcal{D}_{\{\hat{u}\},\{\bar{u}\}} = \mathcal{D}$$

and hence

(5.5.13)
$$\hat{L}_T^{Y,\mathcal{D}\{\hat{u}\},\{\bar{u}\}} = \hat{L}_T^{Y} , \quad \hat{\ell}_T^{Y,\mathcal{D}\{\hat{u}\},\{\bar{u}\}} = \hat{\ell}_T^{Y} ,$$
$$L_T^{Y,\mathcal{D}\{\hat{u}\},\{\bar{u}\},\underline{\epsilon}} = L_T^{Y,\epsilon} , \quad \ell_T^{Y,\mathcal{D}\{\hat{u}\},\{\bar{u}\},\underline{\epsilon}} = \ell_t^{T,\epsilon} .$$

<u>Theorem</u> 5.5.4. Let $\mathcal{X}_0,\mathcal{X}_t,Z_t, t=1,2,\ldots,T$ be finite dimensional Euclidean spaces, let the loss function have structure V, and let Y_t be a sufficient statistic for V for which the kernels K_t^Y and $G_t^{Y,V}$ satisfy (3.2.5) and (3.2.9) where K_t and G_t are those of Theorem 2.2.3. If the system has a single selection class in Y, then the system has a single selection class in the identity statistic I (Definition 3.2.7).

Proof: Let

(5.5.14) $$L_t^{I,\underline{\epsilon}}(z^t,u^{t-1};u_t) = L_t^{Y,\underline{\epsilon}}(Y_t(z^t,u^{t-1}),u_t)$$

(5.5.15) $$\ell_t^{I,\underline{\epsilon}}(z^t,u^{t-1}) = \ell_t^{Y,\underline{\epsilon}}(Y_t(z^t,u^{t-1}))$$

(5.5.16) $$\tilde{\ell}_t^{I,\underline{\epsilon}}(z^t,u^{t-1}) = \tilde{\ell}_t^{Y,\underline{\epsilon}}(Y_t(z^t,u^{t-1}))$$

(5.5.17) $$\hat{U}_t^{I,\underline{\epsilon}}(z^t,u^{t-1}) = \hat{U}_t^{Y,\underline{\epsilon}}(Y_t(z^t,u^{t-1})) .$$

It will be shown that the conditions of Definition 5.2.3 hold for these functions. From (5.5.17) and (5.2.21) for Y, (5.2.21) holds for I. Similarly (5.2.22) holds.

Since by assumption the loss function has structure V, it satisfies (3.2.7). Thus the loss function has identity structure where

(5.5.18) $$c_t^{V=I}(x_t,u^{t-1}) = c_t^V(V_t(x_t,u^{t-1}),u_{t-1}) .$$

From Theorem 2.2.3 the kernels G_t and K_t satisfy (2.1.1) and (2.1.2). Thus from Lemma 3.2.4 it may be assumed that K_t^I and $G_t^{I,I}$ satisfy (3.2.12) and (3.2.13). From (5.2.2) of Definition 2.2.1 for Y, (3.2.9)

and Lemma A.1.13, (5.5.18), (3.2.13), and Definition 5.2.1 for I , for all (z^{t+1},u^t)

$$c^Y_{t+1}(Y_{t+1}(z^{t+1},u^t),u_t) = \int_{\mathcal{V}_{t+1}} c^V_{t+1}(v_{t+1},u_t)G^{Y,V}_{t+1}(Y_{t+1}(z^{t+1},u^t),dv_{t+1})$$

$$= \int_{\mathcal{X}_{t+1}} c^V_{t+1}(V_{t+1}(x_{t+1},u_t),u_t)G_{t+1}(z^{t+1},u^t;dx_{t+1})$$

$$(5.5.19) \qquad = \int_{\mathcal{X}_{t+1}} c^{V=I}_{t+1}(x_{t+1},u^t)G_{t+1}(z^{t+1},u^t;dx_{t+1})$$

$$= \int_{\mathcal{X}_{t+1}\times\mathcal{U}^t} c^{V=I}_{t+1}(x_{t+1},u_t)G^{I,I}_{t+1}(z^{t+1},u^t;dx_{t+1},du^t)$$

$$= c^{Y=I}_{t+1}(z^{t+1},u^t) \quad .$$

From (5.5.14), (5.2.23) of Definition 5.2.3 for Y , (3.2.5) and Lemma A.1.13,(5.5.15), (5.5.19), and (3.2.12), for all (z^t,u^t)

$$L^{I,\underline{\varepsilon}}_t(z^t,u^{t-1};u_t) = L^{Y,\underline{\varepsilon}}_t(Y_t(z^t,u^{t-1}),u_t)$$

$$= \int_{\mathcal{Y}_{t+1}} [\ell^{Y,\underline{\varepsilon}}_{t+1}(y_{t+1}) + c^Y_{t+1}(y_{t+1},u_t)]K^Y_t(y_t(z^t,u^{t-1}),u_t;dy_{t+1})$$

$$= \int_{Z_{t+1}} [\ell^{Y,\underline{\varepsilon}}_{t+1}(Y_{t+1}(z^{t+1},u^t)) + c^Y_{t+1}(Y_{t+1}(z^{t+1},u^t),u^t)]K_t(z^t,u^t;dz_{t+1})$$

$$= \int_{Z_{t+1}} [\ell^{I,\underline{\varepsilon}}_{t+1}(z^{t+1},u^t) + c^{Y=I}_{t+1}(z^{t+1},u^t)]K_t(z^t,u^t;dz_{t+1})$$

$$= \int_{Z_{t+1}\times\mathcal{U}^t} [\ell^{I,\underline{\varepsilon}}_{t+1}(z^{t+1},u^t) + c^{Y=I}_{t+1}(z^{t+1},u^t)]K^I_t(z^t,u^t;dz_{t+1},du^t) \quad .$$

Thus (5.2.23) holds for I . Equation (5.2.27) for I follows from (5.5.15), (5.2.27) for Y , (5.5.17) and (5.5.14). Next it will be shown that equations (5.2.24)-(5.2.26) and (5.2.28) hold for I where B_t for I is defined to be the inverse image of B_t for Y . That is,

$$B^I_t = \{(z^t,u^{t-1}) \mid Y_t(z^t,u^{t-1}) \in B^Y_t\} \quad .$$

Let N be the set in $\mathcal{Y}_t$ for which (5.2.24) , (5.2.25), (5.2.26), or (5.2.28) fail to hold. On the set

$$C = \{(z^t,u^{t-1}) \mid Y_t(z^t,u^{t-1}) \notin N\}$$

the equations (5.2.24)-(5.5.26) and (5.2.28) can easily be shown to hold for I . From Definition 3.2.9 for all $\{u\} \in \mathcal{D}$

$$(I^{-1}P)^t_{\{u\}}(C) = P^t_{\{u\}}[\{z^t | (z^t, u^{t-1}(z^{t-1})) \in C\}]$$

$$= P^t_{\{u\}}[\{z^t | Y_t(z^t, u^{t-1}(z^{t-1})) \notin N\}] = (Y^{-1}P)^t_{\{u\}}(N^c) = 1$$

since N is a null set with respect to $(Y^{-1}P)^t_{\{u\}}$.

Thus (5.2.24)-(5.2.26) and (5.2.8) hold a.s. $(I^{-1}P)^t_{\{u\}}$ for all $\{u\} \in \mathcal{D}$. Monotoneity of $L_t^{I,\underline{\varepsilon}^n}$ and $\ell_t^{Y,\underline{\varepsilon}^n}$ follows from monotoneity of $L_t^{Y,\underline{\varepsilon}^n}$ and $\ell^{Y,\underline{\varepsilon}^n}$ and (5.5.14), (5.5.15) .

If the minimum of the function $L_t^{Y,\mathcal{D}_0,\underline{\varepsilon}}(y_t,u_t)$ in u_t is attained a.s. $(Y^{-1}P)^t_{\{u\}}$ for all t , then additional simplications of the theory are possible. Only the single selection class will be considered.

<u>Definition</u> 5.5.4. The system has a <u>single strong selection class in</u> Y provided there exist measurable functions $\hat{\ell}_t^Y(y_t)$, $\hat{L}_t^Y(y_t,u_t)$ and a law $\{\hat{U}^Y\}$ in Y which satisfy for all $\{u\} \in \mathcal{D}$

(5.5.20) $\qquad \{u^{t-1}\} \vee \{\hat{U}^Y\} \in \mathcal{D} \qquad t = 0,1,\ldots,T-1$

(5.5.21) $\qquad \hat{L}_T^Y = \hat{\ell}_T^Y = 0$

(5.5.22) $\qquad \hat{L}_t^Y(y_t,u_t) = \int_{\mathcal{Y}_{t+1}} [c_{t+1}^Y(y_{t+1},u_t) + \hat{\ell}_{t+1}^Y(y_{t+1})] K_t^Y(y_t,u_t;dy_{t+1})$

(5.5.23) $\qquad \hat{L}_t^Y(y_t,\hat{U}_t^Y(y_t)) = \inf_{u_t \in \mathcal{U}_t} \hat{L}_t(y_t,u_t) \qquad$ a.s. $(Y^{-1}P)^t_{\{u\}}$

(5.5.24) $\qquad \ell_t^Y(y_t) = \hat{L}_t^Y(y_t,\hat{U}_t^Y(y_t))$

$t = 0,1,\ldots,T-1$.

<u>Lemma</u> 5.5.2. If the system has a single strong selection class in Y , then it has a single selection class with

(5.5.25) $\qquad \ell_t^{Y,\underline{\varepsilon}}(y_t) = \tilde{\ell}_t^{Y,\underline{\varepsilon}}(y_t) = \hat{\ell}_t^Y(y_t)$

(5.5.26) $\quad L_t^{Y,\underline{\varepsilon}}(y_t,u_t) = \hat{L}_t^Y(y_t,u_t)$

(5.5.27) $\quad \{\hat{U}^{Y,\underline{\varepsilon}}\} = \{\hat{U}^Y\}$

for all $\underline{\varepsilon}$.

Proof: Except for values of y_t for which (5.5.23) fails to hold, $u^* = \hat{U}_t^Y(y_t)$ is a minimum in u_t of $L_t(y_t,u_t)$. Thus

$$B_t = \mathcal{Y}_t \qquad\qquad \text{a.s. } (Y^{-1}P)^t_{\{u\}}$$

and (5.2.24) – (5.2.26) follows from (5.5.23) and (5.5.24) . From (5.5.25) and (5.5.26) $L_t^{Y,\underline{\varepsilon}^n}$ and $\ell_t^{Y,\underline{\varepsilon}^n}$ are independent of $\underline{\varepsilon}^n$ and hence monotone non-increasing in n .

<u>Theorem</u> 5.5.5. If the system has a single strong selection class in Y , then $\{\hat{U}^Y\}$, $\hat{L}_t^Y$, and $\hat{\ell}_t^Y$ satisfy i), ii), and iii) of Theorem 5.5.3 and in addition

iv) $\{\hat{U}^Y\}$ is a universal optimal law

v) $\{\hat{U}^Y\}$ is optimal at $t = 0$.

Proof: From (5.5.25), (5.5.26), (5.2.30), and (5.2.31) the notation $\hat{\ell}_t^Y$, $\hat{L}_t^Y$ is consistent. From Lemma 5.5.2 the system has a single selection class, and it follows that i)-iii) of Theorem 5.5.3 hold. From Theorem 5.5.2 and (5.5.27) $\{u^{t-1}\} \vee \{\hat{U}^Y\}$ is ε-optimal for $\{u\}$ at t for all $\varepsilon > 0$. Thus

$$\hat{F}_t(z^t,\{u\}) \leq E_{\{u^{t-1}\} \vee \{\hat{U}^Y\}}[L|z^t] \leq \hat{F}_t(z^t,\{u\}) + \frac{1}{n} \qquad \text{a.s. } P^t_{\{u\}}$$

for all n . It follows that

$$E_{\{u^{t-1}\} \vee \{\hat{U}^Y\}}[L|z^t] = \hat{F}_t(z^t,\{u\}) \qquad\qquad \text{a.s. } P^t_{\{u\}} \quad ,$$

and hence that $\{u^{t-1}\} \vee \{\hat{U}^Y\}$ is optimal for $\{u\}$ at t . Property v) follows from iv) for $t = 0$.

<u>Theorem</u> 5.5.6. If $\mathcal{D}$ if universal in Y , the spaces $\mathcal{Y}_t, t=1,2,\ldots,T$ are countable, and

(5.5.28) $\mathcal{B}_{\mathcal{Y}_t} = 2^{\mathcal{Y}_t} = \{\text{all subsets of } \mathcal{Y}_t\}$,

then the system has a single selection class.

Proof: At each step define $\hat{U}_t^{Y,\underline{\varepsilon}}(y_t)$ as follows: for y_t fixed if there exists u^* that satisfies

$$L_t^{Y,\underline{\varepsilon}}(y_t,u^*) = \inf_{u_t \in \mathcal{U}_t} L_t^{Y,\underline{\varepsilon}}(y_t,u_t)$$

then let

$$\hat{U}_t^{Y,\underline{\varepsilon}} = u^* \quad ;$$

if no such u^* exists, select u^{**} such that

$$L_t^{Y,\underline{\varepsilon}}(y_t,u^{**}) \leq \inf_{u_t \in \mathcal{U}_t} L_t^{Y,\underline{\varepsilon}}(y_t,u_t) + \varepsilon_t$$

and let

$$\hat{U}_t^{Y,\underline{\varepsilon}}(y_t) = u^{**} \quad .$$

From (5.5.28) all function of y_t are measurable. Defining $\ell_t^{Y,\underline{\varepsilon}}(y_t)$ and $L_{t-1}^{Y,\underline{\varepsilon}}(y_{t-1},u_{t-1})$ by (5.2.27) and (5.2.23), and $\tilde{\ell}_t^{Y,\underline{\varepsilon}}$ by

$$\tilde{\ell}_t^{Y,\underline{\varepsilon}}(y_t) = \inf_{u_t \in \mathcal{U}_t} L_t^{Y,\underline{\varepsilon}}(y_t,u_t) \quad ,$$

the conditions (5.2.24)-(5.2.28) are clearly satisfied for all y_t. The monotoneity of $\ell_t^{Y,\underline{\varepsilon}^n}$ and $L_t^{Y,\underline{\varepsilon}^n}$ can be assured in the same manner as in the proof of Theorem 5.4.8. It follows then that $\mathcal{D}$ is a selection class.

<u>Lemma</u> 5.5.3. Let $L(y,u)$ be non-negative (possibly $+\infty$) and measurable in y for all $u \in \mathcal{U}$ where the space $\mathcal{U}$ is countable. Then for $\varepsilon > 0$ there exists $U^\varepsilon(y)$ and $\tilde{\ell}(y)$ measurable which satisfy

(5.5.29) $$\tilde{\ell}(y) = \inf_{u \in \mathcal{U}} L(y,u)$$

(5.5.30) $$\chi_B(y)L(y,U^\varepsilon(y) = \chi_B(y)\tilde{\ell}(y)$$

(5.5.31) $$\chi_{B^c}(y)L(y,U^\varepsilon(y)) \leq \chi_{B^c}(y)\tilde{\ell}(y) + \varepsilon$$

for all $y \in \mathcal{Y}$ where

(5.5.32) $\quad B = \{y | \exists u^* \in \mathcal{U} \ni L(y,u^*) = \tilde{\ell}(y)\}$

and $B \in \mathcal{B}_{\mathcal{Y}}$. If $\mathcal{U}$ is finite, then

(5.5.33) $\quad B = \mathcal{Y}$.

Proof: Since $\mathcal{U}$ is countable, $\tilde{\ell}(y)$, defined by (5.5.29) is measurable. Let

$$\mathcal{U} = \{u_n\}_{n=1,2,\ldots}$$

and define

$$B_n = \{y | L(y,u_n) = \tilde{\ell}(y) \text{ and } L(y,u_i) \neq \tilde{\ell}(y)$$

$$\text{for } i = 1,2,\ldots,n-1\} ,$$

(5.5.34) $$B = \bigcup_{n=1}^{\infty} B_n ,$$

$$C_n = \{y | y \notin B , L(y,u_n) \leq \tilde{\ell}(y) + \epsilon$$

$$\text{and } L(y,u_i) > \tilde{\ell}(y) + \epsilon \text{ for } i = 1,\ldots,n-1\} ,$$

and

(5.5.35) $$U^{\epsilon}(y) = \sum_{n=1}^{\infty} u_n \chi_{C_n \cup B_n}(y) .$$

It can easily be shown that the sets $\{B_n, C_n\}$ $n = 1,2,\ldots$ are disjoint, satisfy

$$\sum_{n=1}^{\infty} B_n + \sum_{n=1}^{\infty} C_n = \mathcal{Y} ,$$

that B defined by (5.5.34) satisfies (5.5.32), and that $U^{\epsilon}(y)$ defined by (5.5.35) is measurable and satisfies (5.5.30) and (5.5.31). For $\mathcal{U}$ finite the infimum (5.5.29) is always attained. Thus (5.5.33) holds.

<u>Theorem</u> 5.5.7. Let $\mathcal{D}$ be universal in Y and let the spaces $\mathcal{U}_t$ be countable for $t = 0,1,\ldots,T-1$. Then the system has a single selection class. If each $\mathcal{U}_t$ is finite, then the system has a single strong selection class. Further the laws $\{\hat{U}^{Y,\underline{\epsilon}}\}$ can be constructed by backward induction from (5.4.26)-(5.4.34) where all equations hold for all y_t rather than a.e. ρ_t .

Proof: The functions $\ell_t^{Y,\underline{\varepsilon}}$, $\tilde{\ell}_t^{Y,\underline{\varepsilon}}$, $L_t^{Y,\underline{\varepsilon}}$ and $\hat{U}_t^{Y,\underline{\varepsilon}}$ can be constructed by backward induction as in proof of Theorem 5.4.8, with Lemma 5.5.3 used in place of Lemma 5.4.1 . If the sets $\mathcal{U}_t$ are finite, from Lemma 5.5.3 for all $y \in \mathcal{Y}$ and $\varepsilon > 0$

$$L(y,U^{\varepsilon}(y)) = \tilde{\ell}(y) .$$

Thus in this case from (5.5.29), the law obtained by backward induction also satisfies (5.5.23) of Definition 5.5.4 for all y_t .

Chapter 6 - Quadratic Loss

6.1 Optimal control for the Linear Gaussian model

The classical model for the stochastic control problem is the linear Gaussian model with quadratic loss function. This problem has been studied by many authors. For example, Aoki [1] p. 44 gives a derivation of the optimal control law for this system that is very similar to that of section 6.2. The result of section 6.2 is given primarily as an example of the application of the earlier theory. It is a generalization of existing results only in the rather uninteresting respect that the quadratic forms of the loss function are permitted to be singular.

In section 6.3 the sufficient statistic $\hat{a}_t$ (3.4.14) is used to obtain the optimal law for a quadric loss function with structor a_t (3.4.17). In the applications of sections 6.2 and 6.3, the existence theorems of section 5.5 are not used though they do, in fact, apply. Rather it is shown directly that the system has a single strong selection class. Some existence theorems for the Gaussian model will be stated here although they are not required for alter application.

Theorem 6.1.1. For the linear Gaussian model with loss function

$$(6.1.1) \qquad L = \sum_{t=1}^{T} c_t^V(x_t, \sum_{\tau=0}^{t-1} \bar{c}_\tau(u_\tau), u_{t-1})$$

where the $\bar{c}_\tau$ are measurable functions taking values in a finite dimensional Euclidean space, let $\mathfrak{D}$ be universal in the statistic

$$(6.1.2) \qquad Y_t = (\hat{x}_t, \sum_{\tau=0}^{t-1} \bar{c}_\tau(u_\tau)) .$$

Then for t, $\{u\} \in \mathfrak{D}$, and $\epsilon > 0$, there exists a control law $\{\hat{U}^{Y,\epsilon}\}$ with the properties

i) $\{u^{t-1}\} \vee \{\hat{U}^{Y,\epsilon}\}$ is ϵ-optimal for $\{u\}$ at t, and

ii) if there exists $\{\hat{u}\}$ optimal for $\{u\}$ at t, then $\{u^{t-1}\} \vee \{\hat{U}^{Y,\epsilon}\}$ is optimal for $\{u\}$ at t.

Proof: From Theorem 3.4.4 i) Y_t is sufficient for the loss function (6.1.1). For the linear Gaussian system the statistic Y_t given by (6.1.2) and the control functions u_t take values in finite dimensional Euclidean spaces and hence complete separable metric spaces. Thus Corollary 5.4.1 applies, and the system has countably complete selection classes in Y. The result then follows from Theorem 5.4.3.

<u>Theorem</u> 6.1.2. For the linear Gaussian model with loss function

$$L = \sum_{t=1}^{T} c_t^V(A_T \Phi_{T,t} x_t, \sum_{\tau=0}^{t-1} \bar{c}_\tau(u_\tau), u_{t-1}) \quad , \tag{6.1.3}$$

let $\mathfrak{D}$ be universal in the statistic

$$Y_t = (\hat{a}_t , \sum_{\tau=0}^{t-1} \bar{c}_\tau(u_\tau)) \quad , \tag{6.1.4}$$

let the $\bar{c}_\tau$ take values in a countable subspace, and let the covariances $\bar{\Phi}_t$, $t=0,1,\ldots,T-1$ given by (3.4.20) be non-singular. Then there is a single selection class, and for $\epsilon > 0$ there exists a control law $\{\hat{U}^{T,c}\}$ that is universally ϵ-optimal. Further $\{\hat{U}^{Y,\epsilon}\}$ can be constructed by the backward iteration (5.4.26)-(5.4.34) where

$$\epsilon = \sum_{\tau=0}^{T-1} \epsilon_\tau \tag{6.1.5}$$

and ρ_t is the product of Lebesgue measure on the space of $\hat{a}_t$ and the counting measure on the space of

$$m_{t-1} = \sum_{\tau=0}^{t-1} \bar{c}_\tau(u_\tau) \quad . \tag{6.1.6}$$

Proof: From Theorem 3.4.4 ii) Y_t is a sufficient statistic. From (3.4.18) of Theorem 3.4.3 and Lemma 3.2.5 the product measure ρ_t dominates the statistic Y_t . For the linear Gaussian model the spaces $\mathcal{U}_t$ are finite dimensional Euclidean spaces and hence complete separable metric spaces. From the assumptions of the theorem $\mathcal{Y}_t$ is the product of finite dimensional Euclidean vector space and a countable space, and hence is a complete separable metric space. Thus from Theorem 5.5.1 there is a single selection class, and the laws $\{\hat{U}^{Y,\epsilon}\}$ can be constructed by (5.4.26)-(5.4.34). From Theorem 5.5.2 these laws are universally ϵ-optimal where ϵ is given by (6.1.5) .

A few comments are in order concerning the matrix terminology which will be followed in this chapter. A matrix Q is said to be positive semi-definite provided it is symmetric and satisfies

(6.1.7) $$x'Qx \geq 0$$

for all real column vectors x. The trace of a matrix Σ is the sum of the diagonal elements

(6.1.8) $$\operatorname{tr} \Sigma = \sum_i \sigma_{ii} \quad .$$

6.2 General quadratic loss

For the linear Gaussian system consider the loss function

(6.2.1) $$L = \sum_{t=1}^{T} (x_t' Q_t^* x_t + u_{t-1}' Q_{t-1} u_{t-1})$$

where Q_t^* and Q_t are positive semi-definite matrices. From Theorem 3.4.2, the statistic $\hat{x}_t$ given by (2.4.5), (2.4.6) is sufficient for the structure $V_t = x_t$ with control kernels $K_t^{\hat{x}}$ and $G_t^{\hat{x},x}$ given by (3.4.6) and (3.4.8). The loss function (6.2.1) satisfies (3.2.7) of Definition 3.2.5 for $V_t = x_t$ with

(6.2.2) $$c_t^x(x_t, u_{t-1}) = x_t' Q_t^* x_t + u_{t-1}' Q_{t-1} u_{t-1} \quad .$$

Thus (6.2.1) has structure $V_t = x_t$.

A lemma concerning minimization of quadratic forms will be required for the computation of an optimum control for the loss function of this and the next section.

Lemma 6.2.1. Let ζ, Q_1, and Q_2 be $r \times s$, $s \times s$, and $r \times r$ matrices where Q_1 and Q_2 are positive semi-definite. Then there exists an $s \times r$ matrix M such that

(6.2.3) $$(Q_1 + \zeta' Q_2 \zeta) M = -\zeta' Q_2 \quad .$$

Further

(6.2.4) $\hat{u} = M\alpha$

satisfies

(6.2.5)
$$\hat{u}'(Q_1 + \zeta'Q_2\zeta)\hat{u} + \hat{u}'\zeta'Q_2\alpha + \alpha'Q_2\zeta\hat{u}$$
$$= \inf_u [u'(Q_1 + \zeta'Q_2\zeta)u + u'\zeta'Q_2\alpha + \alpha'Q_2\zeta u]$$
$$= \alpha'(I+\zeta M)'Q_2(I+\zeta M)\alpha - \alpha'Q_2\alpha + \alpha'M'Q_1M\alpha$$

for all r dimensional column vectors α .

Proof: The existence of the matrix M will be demonstrated first. Since Q_1 and Q_2 are positive semi-definite, the matrix $Q_1 + \zeta'Q_2\zeta$ is positive semi-definite and hence can be diagonalized by

(6.2.6) $Q_1 + \zeta'Q_2\zeta = E\,\Xi\,E'$

where

$$E = [e_1, e_2, \ldots, e_s]$$

is an orthogonal matrix of eigenvectors and

$$\Xi = \begin{bmatrix} \xi_1, & & & 0 \\ & \xi_2, & & \\ & & \ddots & \\ 0 & & & ,\xi_s \end{bmatrix}$$

is a diagonal matrix of eigenvalues. Thus

(6.2.7) $(Q_1 + \zeta'Q_2\zeta)e_i = \xi_i e_i \qquad i = 1,2,\ldots,s$.

From (6.2.6) M satisfies (6.2.3) if and only if

(6.2.8) $\Xi E'M = -E'\zeta'Q_2$.

Letting

(6.2.9) $B = E'M$ and $C = -E'\zeta'Q_2$,

(6.2.8) becomes

(6.2.10) $\xi_i B_i = -C_i \qquad i = 1,2,\ldots,s$.

where B_i , C_i are the row vectors of the matrices B and C . In order to show that a solution B_i exists to the equation (6.2.10),it must be shown that $\xi_i = 0$ implies that $C_i = 0$. Thus, let $\xi_i = 0$. From (6.2.7)

$$(6.2.11) \qquad e_i'(Q_1 + \zeta' Q_2 \zeta) e_i = 0 .$$

Since Q_1 and Q_2 are positive semi-definite,

$$e_i' Q_1 e_i \geq 0$$

and

$$e_i' \zeta' Q_2 \zeta e_i \geq 0 .$$

Thus (6.2.11) implies that

$$(6.2.12) \qquad e_i' \zeta' Q_2 \zeta e_i = 0 .$$

Since Q_2 is positive semi-definite, there exists Q_3 positive definite and F not necessarily square such that

$$(6.2.13) \qquad Q_2 = F' Q_3 F .$$

Thus from (6.2.12)

$$e_i' \zeta' F' Q_3 F \zeta e_i = 0 .$$

Since Q_3 is positive definite, this implies that

$$e_i' \zeta' F' = 0$$

and hence from (6.2.9) and (6.2.13)

$$C_i = -e_i' \zeta' Q_2 = -(e_i' \zeta' F') Q_3 F = 0 .$$

Thus a solution to (6.2.10)is obtained by

$$B_i = -(\frac{1}{\xi_i}) C_i$$

for $\xi_i \neq 0$ and B_i arbitrary for $\xi_i = 0$. Letting

$$M = EB$$

a solution to (6.2.3) is obtained. From (6.2.3)

$$u'(Q_1 + \zeta' Q_2 \zeta)u + u'\zeta' Q_2 \alpha + \alpha' Q_2 \zeta u$$

(6.2.14)
$$= u'(Q_1 + \zeta' Q_2 \zeta)u - u'(Q_1 + \zeta' Q_2 \zeta)M\alpha - \alpha' M'(Q_1 + \zeta' Q_2)u$$

$$= (u - M\alpha)'(Q_1 + \zeta' Q_2 \zeta)(u - M\alpha) - \alpha' M'(Q_1 + \zeta' Q_2 \zeta)M\alpha \; .$$

Since $Q_1 + \zeta' Q_2 \zeta$ is positive semi-definite, the first term

$$(u - M\alpha)'(Q_1 + \zeta' Q_2 \zeta)(u - M\alpha) \geq 0$$

for all u. Thus the minimum of the quadratic form (6.2.14) is obtained by

$$\hat{u} = M\alpha \quad ,$$

and the first equality in (6.2.5) follows. The second equality follows from (6.2.3) and (6.2.14).

Theorem 6.2.1. For the linear Gaussian model (1.3.1)-(1.3.5) and the loss function (6.2.1) where the matrices Q_t and Q_t^* are positive semi-definite, there exist matrices $\hat{M}_t$ and $\mathfrak{Q}_t$ which satisfy

(6.2.15)
$$\mathfrak{Q}_T = 0$$

(6.2.16)
$$[Q_t + \Lambda_t'(Q_{t+1}^* + \mathfrak{Q}_{t+1})\Lambda_t]\hat{M}_t = -\Lambda_t'(Q_{t+1}^* + \mathfrak{Q}_{t+1})$$

(6.2.17)
$$\mathfrak{Q}_t = \Phi_t'(I + \Lambda_t \hat{M}_t)'(Q_{t+1}^* + \mathfrak{Q}_{t+1})(I + \Lambda_t \hat{M}_t)\Phi_t + \Phi_t' \hat{M}_t' Q_t \hat{M}_t \Phi_t$$

$t = T-1, T-2, \ldots, 0$; and the system has a single strong selection class in $\hat{x}_t$ with

(6.2.18)
$$\hat{U}_t^{\hat{x}}(\hat{x}_t) = \hat{M}_t \Phi_t \hat{x}_t \quad ,$$

(6.2.19)
$$\mathfrak{L}_t^{\hat{x}}(\hat{x}_t) = \hat{x}_t' \mathfrak{Q}_t \hat{x}_t + \sum_{\tau=t+1}^{T} \operatorname{tr}(Q_\tau^* \Pi_\tau + Q_\tau^* \mathfrak{F}_\tau + \mathfrak{Q}_\tau \mathfrak{F}_\tau)$$

(6.2.20)
$$L_t^{\hat{x}}(\hat{x}_t, u_t) = (\Phi_t \hat{x}_t + \Lambda_t u_t)'(Q_{t+1}^* + \mathfrak{Q}_{t+1})(\Phi_t \hat{x}_t + \Lambda_t u_t) + u_t' Q_t u_t + \sum_{\tau=t+1}^{T} \operatorname{tr}(Q_\tau^* \Pi_\tau + Q_\tau^* \mathfrak{F}_\tau + \mathfrak{Q}_\tau \mathfrak{F}_\tau)$$

provided $\{u^{t-1}\} \vee \{\hat{U}^{\hat{x}}\} \in \mathfrak{D}$ for all $\{u\} \in \mathfrak{D}$. (Π_t and $\mathfrak{F}_t$ are given by (2.4.6) and (3.4.7)).

Proof: It can be shown by backward induction that there exists $\hat{M}_t$ which satisfies (6.2.16) and that $\mathfrak{D}_t$ defined by (6.2.17) is positive semi-definite. Existence of $\hat{M}_t$ follows from Lemma 6.2.1 with $Q_1 = Q_t$, $\zeta = \Lambda_t$, and $Q_2 = Q^*_{t+1} + \mathfrak{D}_{t+1}$. That $\mathfrak{D}_t$ is positive semi-definite follows from its form (6.2.17) and the fact that $Q^*_{t+1}, \mathfrak{D}_{t+1}$, and Q_t are positive semi-defininte.

It will be shown that $\{\hat{U}^{\hat{x}}\}$, $\hat{\ell}^{\hat{x}}_t$, and $\hat{L}^{\hat{x}}_t$ defined by (6.2.18)-(6.2.20) satisfy (5.5.20)-(5.5.24) of Definition 5.5.4. By assumption (5.5.20) holds. For $t = T$, from (6.2.15) and (6.2.19), $\hat{\ell}^{\hat{x}}_T = 0$, and by convention $\hat{L}^{\hat{x}}_T = 0$. Thus (5.5.21) holds. Next, $c^{\hat{x}}_{t+1}$ defined by (5.2.2) will be computed. From (5.2.2), (6.2.2), (3.4.8), and the properties of the normal distribution

(6.2.21)
$$c^{\hat{x}}_{t+1}(\hat{x}_{t+1}, u_t) = \int_{\mathcal{X}_{t+1}} (x'_{t+1} Q^*_{t+1} x_{t+1} + u'_t Q_t u_t) G^{\hat{x},x}_{t+1}(\hat{x}_{t+1}, dx_{t+1})$$
$$= \hat{x}'_{t+1} Q^*_{t+1} \hat{x}_{t+1} + u'_t Q_t u_t + \mathrm{tr}(Q^*_{t+1} \Pi_{t+1}) \quad .$$

Next it will be shown that $\hat{L}^{\hat{x}}_t$ given by (6.2.20) satisfies (5.5.22). From (6.2.19), (6.2.21), (6.2.19), and (3.4.6)

$$\int_{\mathcal{X}_{t+1}} [c^{\hat{x}}_{t+1}(\hat{x}_{t+1}, u_t) + \hat{\ell}^{\hat{x}}_{t+1}(\hat{x}_{t+1})] K^{\hat{x}}_t(\hat{x}_t, u_t; d\hat{x}_{t+1})$$
$$= \int_{\mathcal{X}_{t+1}} [\hat{x}'_{t+1}(\mathfrak{D}_{t+1} + Q^*_{t+1})\hat{x}_{t+1} + u'_t Q_t u_t + \mathrm{tr}(Q^*_{t+1}\Pi_{t+1})$$
$$+ \sum_{\tau=t+2}^{T} \mathrm{tr}(Q^*_\tau \Pi_\tau + Q^*_\tau \Sigma_\tau + \mathfrak{D}_\tau \Sigma_\tau)] K^{\hat{x}}_t(\hat{x}_t, u_t; d\hat{x}_{t+1})$$
$$= (\Phi_t \hat{x}_t + \Lambda_t u_t)'(\mathfrak{D}_{t+1} + Q^*_{t+1})(\Phi_t \hat{x}_t + \Lambda_t u_t) + u'_t Q_t u_t$$
$$+ \sum_{\tau=t+1}^{T} \mathrm{tr}(Q^*_\tau \Pi_\tau + Q^*_\tau \Sigma_\tau + \mathfrak{D}_\tau \Sigma_\tau) = \hat{L}^{\hat{x}}_t(\hat{x}_t, u_t) \quad .$$

Thus (5.5.22) is satisfied for all t. From (6.2.20)

$$\hat{L}^{\hat{x}}_t(\hat{x}_t, u_t) = u'_t[\Lambda'_t(\mathfrak{D}_{t+1} + Q'_{t+1})\Lambda_t + Q_t]u_t$$
$$+ (\Phi_t \hat{x}_t)'(\mathfrak{D}_{t+1} + Q^*_{t+1})\Lambda_t u_t + u'_t \Lambda'_t(\mathfrak{D}_{t+1} + Q^*_{t+1})(\Phi_t \hat{x}_t)$$
$$+ \hat{x}'_t \Phi'_t(\mathfrak{D}_{t+1} + Q^*_{t+1})\Phi_t \hat{x}_t + \sum_{\tau=t+1}^{T} \mathrm{tr}(Q^*_\tau \Pi_\tau + Q^*_\tau \Sigma_\tau + \mathfrak{D}_\tau \Sigma_\tau) \quad .$$

Thus from (6.2.18), (6.2.4) of Lemma 6.2.1 with $\alpha = \Phi_t\hat{x}_t$, (6.2.17), and (6.2.19)

$$\hat{L}_t^{\hat{x}}(\hat{x}_t,\hat{U}_t^{\hat{x}}(\hat{x}_t)) = \inf_{u_t} \{u_t'[Q_t + \Lambda_t'(Q_{t+1}^* + \mathfrak{D}_{t+1})\Lambda_t]u_t$$

$$+ u_t'\zeta_t'(Q_{t+1}^* + \mathfrak{D}_{t+1})(\Phi_t\hat{x}_t) + (\Phi_t\hat{x}_t)'(Q_{t+1}^* + \mathfrak{D}_{t+1})\Lambda_t u_t\}$$

$$+ \hat{x}_t'\Phi_t'(\mathfrak{D}_{t+1} + Q_{t+1}^*)\Phi_t\hat{x}_t + \sum_{\tau=t+1}^{T} \operatorname{tr}(Q_\tau^*\Pi_\tau + Q_\tau^*\Sigma_\tau + \mathfrak{D}_\tau\Sigma_\tau)$$

$$= \inf_{u_t} \hat{L}_t^{\hat{x}}(\hat{x}_t,u_t) = \hat{x}_t'\Phi_t'(I + \Lambda_t\hat{M}_t)'(Q_{t+1}^* + \mathfrak{D}_{t+1})(I + \Lambda_t\hat{M}_t)\Phi_t\hat{x}_t$$

$$- \hat{x}_t'\Phi_t'(Q_{t+1}^* + \mathfrak{D}_{t+1})\Phi_t\hat{x}_t + \hat{x}_t'\Phi_t'\hat{M}_t'Q_t\hat{M}_t\Phi_t\hat{x}_t$$

$$+ \hat{x}_t'\Phi_t'(\mathfrak{D}_{t+1} + Q_{t+1}^*)\Phi_t\hat{x}_t + \sum_{\tau=t+1}^{T} \operatorname{tr}(Q_\tau^*\Pi_\tau + Q_\tau^*\Sigma_\tau + \mathfrak{D}_\tau\Sigma_\tau)$$

$$= \hat{x}_t'\mathfrak{D}_t\hat{x}_t + \sum_{\tau=t+1}^{T} \operatorname{tr}(Q_\tau^*\Pi_\tau + Q_\tau^*\Sigma_\tau + \mathfrak{D}_\tau\Sigma_\tau)$$

$$= \ell_t^{\hat{x}}(\hat{x}_t) \quad .$$

Thus (5.5,23) and (5.5.24) hold for all t . The result follows.

6.3 Quadratic final miss loss function

In this section an optimum control law in the sufficient statistic $\hat{\alpha}_t$ (3.4.14) will be derived for the linear Gaussian model and the quadratic loss function

$$(6.3.1) \qquad L(x^T,u^{T-1}) = (A_Tx_T)'\, Q_T^*(A_Tx_T) + \sum_{t=0}^{T-1} u_t'Q_tu_t \quad .$$

The vector A_Tx_T represents an error at the final time T and will be called the <u>final miss</u>. Following the discussion in Section 3.4., $\hat{\alpha}_t$ defined by (3.4.14) is the <u>expected final miss</u>.

This loss function has the form (3.2.7) for

$$(6.3.2) \qquad a_t = A_T\Phi_{T,t}x_t \quad ,$$

$$(6.3.3) \qquad c_T^a(a_T,u_{T-1}) = a_T'Q_T^*a_T + u_{T-1}'Q_{T-1}u_{T-1} \quad ,$$

and

$$(6.3.4) \qquad c_t^a(a_t,u_{t-1}) = u_{t-1}'Q_{t-1}u_{t-1}$$

for $t = 1,2,\ldots,T-1$. From Theorem 3.4.3, $\hat{a}_t$ defined by (3.4.14) is sufficient for the structure a_t (6.3.2), and the control and filtering distributions $K_t^{\hat{a}}$ and $G_t^{\hat{a},a}$ are given by (3.4.18) and (3.4.21).

Theorem 6.3.1. For the linear Gaussian model (1.3.1)-(1.3.5) and loss function (6.3.1) where the matrices Q_T^* and Q_t for $t = 0,1,\ldots,T-1$ are positive semi-definite, there exist matrices $\hat{M}_t$ and $\mathfrak{D}_t$ which satisfy

(6.3.5) $$\mathfrak{D}_T = Q_T^*$$

(6.3.6) $$(Q_t + \zeta_t'\mathfrak{D}_{t+1}\zeta_t)\hat{M}_t = -\zeta_t'\mathfrak{D}_{t+1}$$

(6.3.7) $$\mathfrak{D}_t = \hat{M}_t'Q_t\hat{M}_t + (I+\hat{M}_t'\zeta_t')\mathfrak{D}_{t+1}(I+\zeta_t\hat{M}_t)$$

$t = T-1, T-2,\ldots,1,0$; and the system has a single strong selection class in $\hat{a}_t$ with

(6.3.8) $$\hat{U}_t^{\hat{a}}(\hat{a}_t) = \hat{M}_t\hat{a}_t$$

(6.3.9) $$\hat{\ell}_t^{\hat{a}}(\hat{a}_t) = \hat{a}_t'\mathfrak{D}_t\hat{a}_t + \mathrm{tr}(Q_T^*A_T\Pi_TA_T') + \sum_{\tau=t+1}^{T} \mathrm{tr}(\mathfrak{D}_\tau\bar{\bar{\Phi}}_\tau)$$

(6.3.10) $$\hat{L}_t^{\hat{a}}(\hat{a}_t,u_t) = (\hat{a}_t+\zeta_tu_t)'\mathfrak{D}_{t+1}(\hat{a}_t+\zeta_tu_t) + u_t'Q_tu_t$$
$$\mathrm{tr}(Q_T^*A_T\Pi_TA_T') + \sum_{\tau=t+1}^{T} (\mathfrak{D}_\tau\bar{\bar{\Phi}}_\tau)$$

$t = 0,1,\ldots,T-1$, provided $\{u^{t-1}\} \vee \{\hat{U}^{\hat{a}}\} \in \mathfrak{D}$ for all $\{u\} \in \mathfrak{D}$. (ζ_t, Π_t , and $\bar{\bar{\Phi}}_t$ are given by (3.4.19), (2.4.6), and (3.4.20)).

Proof: It can be shown by backward induction that there exists $\hat{M}_t$ which satisfies (6.3.6) and that $\mathfrak{D}_t$ defined by (6.3.7) is positive semi-definite. Existence of $\hat{M}_t$ follows from Lemma 6.2.1 with $Q_1 = Q_t$, $\zeta = \zeta_t$, and $Q_2 = \mathfrak{D}_{t+1}$. That $\mathfrak{D}_t$ is positive semi-definite follows from its form (6.3.7) and the fact that Q_t and $\mathfrak{D}_{t+1}$ are positive semi-definite.

It will be shown that $\{\hat{U}^{\hat{a}}\}$, $\hat{\ell}_t^{\hat{a}}$ and $\hat{L}_t^{\hat{a}}$ defined by (6.3.38)-(6.3.10) satisfy (5.5.20)-(5.5.24) of Definition 5.5.4. It is assumed that (5.5.20) holds, and it can be assumed that (5.5.21) holds. First, the cost functions c_t^a will be computed. From (5.2.2), (6.3.3), (3.4.21), (3.4.13), and properties of the normal distribution

$$(6.3.11)\qquad c_T^{\hat{a}}(\hat{a}_T,u_{T-1}) = \int_{\mathcal{V}_T} [a_T'Q_T^*a_T + u_{T-1}'Q_{T-1}u_{T-1}]G_T^{\hat{a},a}(\hat{a}_T,da_T)$$

$$= \hat{a}_T'Q_T^*\hat{a}_T - u_{T-1}'Q_{T-1}u_{T-1} + \mathrm{tr}(Q_T^*A_T\Pi_TA_T') ,$$

and for $t = 0,1,\ldots,T-2$ from (6.3.4)

$$(6.3.12)\qquad c_{t+1}^{\hat{a}}(\hat{a}_{t+1},u_t) = \int_{\mathcal{V}_{t+1}} (u_t'Q_tu_t)G_{t+1}^{\hat{a},a}(\hat{a}_{t+1},da_{t+1})$$

$$= u_t'Q_tu_t \quad .$$

From (5.5.21), (6.3.11), (3.4.18), (6.3.10) for $t = T-1$, and (6.3.5)

$$\int[c_T^{\hat{a}}(\hat{a}_T,u_{T-1}) + \hat{\ell}_T^{\hat{a}}]K_{T-1}^{\hat{a}}(\hat{a}_{T-1},u_{T-1};d\hat{a}_T)$$

$$= \int[\hat{a}_T'Q_T^*\hat{a}_T + u_{T-1}'Q_{T-1}u_{T-1} + \mathrm{tr}(Q_T^*A_T\Pi_TA_T')]K_{T-1}^{\hat{a}}(\hat{a}_{T-1},u_{T-1};d\hat{a}_T)$$

$$= (\hat{a}_{T-1} + \zeta_{T-1}u_{T-1})'Q_T^*(\hat{a}_{T-1} + \zeta_{T-1}u_{T-1}) + u_{T-1}'Q_{T-1}M_{T-1}$$

$$+ \mathrm{tr}(Q_T^*\bar{\bar{\Phi}}_T + Q_T^*A_T\Pi_TA_T') = L_{T-1}(\hat{a}_{T-1},u_{T-1}) \quad .$$

Thus (5.5.22) holds for $T-1$. From (6.3.12), (6.3.9), (3.4.18), and (6.3.10) for $t < T-1$

$$\int[c_{t+1}^{\hat{a}}(\hat{a}_{t+1},u_t) + \hat{\ell}_{t+1}^{\hat{a}}(\hat{a}_{t+1})]K_t^{\hat{a}}(\hat{a}_t,u_t;d\hat{a}_{t+1})$$

$$= \int [u_t'Q_tu_t + \hat{a}_{t+1}'\mathfrak{D}_{t+1}\hat{a}_{t+1} + \mathrm{tr}(Q_T^*A_T\Pi_TA_T')$$

$$+ \sum_{\tau=t+2}^{T} \mathrm{tr}(\mathfrak{D}_\tau\bar{\bar{\Phi}}_\tau)]K_t^{\hat{a}}(\hat{a}_t,u_t;d\hat{a}_{t+1})$$

$$= u_t'Q_tu_t + (\hat{a}_t + \zeta_tu_t)'\mathfrak{D}_{t+1}(\hat{a}_t + \zeta_tu_t) + \mathrm{tr}(\mathfrak{D}_{t+1}\bar{\bar{\Phi}}_{t+1})$$

$$+ \mathrm{tr}(Q_T^*A_T\Pi_TA_T') + \sum_{\tau=t+2}^{T} \mathrm{tr}(\mathfrak{D}_\tau\bar{\bar{\Phi}}_\tau) = \hat{L}^{\hat{a}}(\hat{a}_t,u_t) \quad .$$

Thus (5.5.22) holds for all t . From (6.3.10)

$$\hat{L}_t^{\hat{a}}(\hat{a}_t,u_t) = u_t'(\zeta_t'\mathfrak{D}_{t+1}\zeta_t + Q_t)u_t + u_t'\zeta_t'\mathfrak{D}_{t+1}\hat{a}_t + \hat{a}_t'\mathfrak{D}_{t+1}\zeta_tu_t$$

$$+ \hat{a}_t'\mathfrak{D}_{t+1}\hat{a}_t + \mathrm{tr}(Q_T^*A_T\Pi_TA_T') + \sum_{\tau=t+1}^{T} \mathrm{tr}(\mathfrak{D}_\tau\bar{\bar{\Phi}}_\tau) \ .$$

Thus from (6.2.4) of Lemma 6.2.1 for $\hat{U}_t^{\hat{a}}$ given by (6.3.8), (6.3.7), and (6.3.9)

$$\hat{L}_t(\hat{a}_t, \hat{U}_t^{\hat{a}}(\hat{a}_t)) = \inf_{u_t} [u_t'(\zeta_t'\mathfrak{D}_{t+1}\zeta_t + Q_t)u_t + u_t'\zeta_t'\mathfrak{D}_{t+1}\hat{a}_t$$

$$+ \hat{a}_t'\mathfrak{D}_{t+1}\zeta_t u_t] + \hat{a}_t'\mathfrak{D}_{t+1}\hat{a}_t + tr(Q_T^* A_T \Pi_T A_T') + \sum_{\tau=t+1}^{T} tr(\mathfrak{D}_\tau \bar{\Sigma}_\tau)$$

$$= \inf_{u_t} \hat{L}_t^{\hat{a}}(\hat{a}_t, u_t) = \hat{a}_t'(I + \zeta_t\hat{M}_t)'\mathfrak{D}_{t+1}(I + \zeta_t\hat{M}_t)\hat{a}_t$$

$$- \hat{a}_t\mathfrak{D}_{t+1}\hat{a}_t + \hat{a}_t'\hat{M}_t'Q_t\hat{M}_t\hat{a}_t + \hat{a}_t\mathfrak{D}_{t+1}\hat{a}_t + tr(Q_T^* A_T \Pi_T A_T') + \sum_{\tau=t+1}^{T} tr(\mathfrak{D}_\tau \bar{\Sigma}_\tau)$$

$$= \hat{a}_t'\mathfrak{D}_t\hat{a}_t + tr(Q_T^* A_T \Pi_T A_T') + \sum_{\tau=t+1}^{T} tr(\mathfrak{D}_\tau \bar{\Sigma}_\tau) = \ell_t^{\hat{a}}(\hat{a}_t) \quad .$$

Thus (5.5.23) and (5.5.24) hold.

Chapter 7. An Absolute Value Loss Function

7.1 Introduction

This chapter is concerned with a linear Gaussian system for which the estimated miss statistic $\hat{a}_t$, defined by (3.4.14), is one dimensional. The loss (7.4.1) for the system combines three types of losses. It is assumed that the system process x_t represents an error or deviation from some desired value. This is the case, for example, in space flight problems for which x_t is obtained by linearizing around a nominal trajectory. A discussion of this procedure and some examples are given by Tung and Striebel [19]. For a model of this type the purpose of control is to keep the error x_t as small as possible. In the loss function (7.4.1) this error is assumed to be of consequence only at the final time T , and loss from this error is measured by the cost function $c(a_T)$. Though only qualitative assumptions will be made concerning the function $c(a_T)$, its typical form is quadratic,

$$(7.1.1) \qquad c(a_T) = a_T^2 \quad .$$

The cost of exerting control is taken into account in the second and third terms of (7.4.1). The second term represents a cost proportional to the absolute value of the control vector, and the third term is a fixed penalty for the use of control regardless of its magnitude. This is intended to apply, for example, when control is applied by the firing of a limited number of rockets, each of which can be burned for a variable length of time but cannot be re-ignited. While the function (7.1.1) is perhaps a reasonable form for the cost function, for this problem, the second two terms of (7.4.1) are not. Certainly, for space flights there is a limited amount of fuel aboard. Thus minimizing the expected amount of fuel consumed is hardly appropriate. A reasonable cost function for this problem is given by (3.1.4) and is discussed in section 3.1. The loss function (7.4.1) provides a not unreasonable approximation to the more difficult problem with loss function of the type (3.1.4). The parameters λ_1 and λ_2 are used to take into account the relative magnitudes of the three types of costs involved.

The absolute value loss function $(\lambda_2 = 0)$ with quadratic $c(a_T)$ (7.1.1) is considered by Tung and Striebel [19] for a continuous time observation model. In this paper, the optimal control law of Theorem 7.4.2 ii) is derived in a slightly hueristic manner, and its performance is compared with several sub-optimal laws.

Sections 7.2 and 7.3 present preliminary results which are used in section 7.4 to obtain optimal control laws. In addition to standard properties of the Gaussian convolution, the principal tool is the variation diminishing property of the convolution given by Karlin [10] .

The results of section 7.4 are all obtained under the assumption of positive observation error variances (7.4.5), (7.4.6) and non-decreasing sensitivity (7.4.13), (7.4.14). The results of the preliminary sections 7.2 and 7.3 can be used to derive optimal control laws without these assumptions. However, the form of the optimal law in these cases becomes considerably more complicated and will not be presented here.

7.2 Continuity and variation diminishing properties of the Gaussian convolution

In this section properties of the convolution

$$N(\alpha) = \int_{-\infty}^{\infty} \nu(\alpha + \eta) n(\eta, \sigma^2) d\eta \tag{7.2.1}$$

will be developed for application in the derivation of an optimal control law for the absolute-value loss function (7.4.1). The notation $n(\eta, \sigma^2)$ indicates the one-dimensional normal density with mean zero and variance σ^2

$$n(\eta, \sigma^2) = \frac{1}{\sqrt{2\pi}\,\sigma} e^{\frac{-\eta^2}{2\sigma^2}} \qquad \sigma > 0 . \tag{7.2.2}$$

Throughout this section σ^2 is fixed and will be omitted from the notation. Thus

$$n(\eta) = n(\eta, \sigma^2) \quad .$$

Variation diminishing properties will be taken from Karlin [10] . Following Karlin, for $\ell(\alpha), -\infty < \alpha < \infty$, define

(7.2.3) $$S^-[\ell(\alpha)] = \sup S^-[\ell(\alpha_1), \ell(\alpha_2), \ldots, \ell(\alpha_m)]$$

(7.2.4) $$S^+[\ell(\alpha)] = \sup S^+[\ell(\alpha_1), \ell(\alpha_2), \ldots, \ell(\alpha_m)]$$

where the supremum is extended over all finite sets $-\infty < \alpha_1 < \alpha_2 < \ldots < \alpha_m < \infty$, $S^-[x_1, x_2, \ldots, x_m]$ is the number of sign changes of the indicated sequence, zero terms being discarded and $S^+[x_1, x_2, \ldots, x_m]$ is the number of sign changes, the zero terms being permitted to take on arbitrary sign. Thus, for example, let

$$\ell(\alpha) = \alpha^3 + \alpha^2$$

then

$$S^-[\ell(\alpha)] = 1 \qquad S^+[\ell(\alpha)] = 3 \quad .$$

First, some results on the sign changes of a function. Obviously from the definition

(7.2.5) $$S^-[\ell] \leq S^+[\ell]$$

and for $1 < i < n$

(7.2.6) $$S^-[x_1, \ldots, x_n] = S^-[x_1, \ldots, x_i] + S^-[x_i, x_{i+1}, \ldots, x_n] \quad .$$

For an interval I, define

(7.2.7) $$S^-_I[\ell] = \sup_{\substack{\alpha_1 < \alpha_2 < \ldots < \alpha_m \\ \alpha_i \in I}} S^-[\ell(\alpha_1), \ell(\alpha_2), \ldots, \ell(\alpha_m)] \quad .$$

By convention, for the degenerate interval

(7.2.8) $$S^-_{[b,b]}[\ell] = 0 \ .$$

The usual conventions for intervals will be followed. Thus, for example, the interval $[a,c)$ contains the lower end point a but not the upper end point c, and $[b,b]$ consists of the single point b.

Lemma 7.2.1. For $b \in I$, an interval,

(7.2.9) $$S^-_I[\ell] = S^-_{I \cap (-\infty,b]}[\ell] + S^-_{I \cap [b,\infty)}[\ell] \quad .$$

Proof: For convenience, let $I = (a,c)$.

Then from the definition (7.2.7) and (7.2.6)

$$S^-_I[\ell] = \sup_{a<\alpha_1<\alpha_2<\dots<\alpha_m<c} S^-[\ell(\alpha_1),\dots,\ell(\alpha_m)] = \sup_{\substack{a<\beta_1<\dots<\beta_n\leq b \\ b\leq\gamma_1<\dots<\gamma_r<c}} S^-[\ell(\beta_1),\dots,\ell(\beta_n),\ell(b),\ell(\gamma_1),\dots,\ell(\gamma_r)]$$

$$= \sup_{a<\beta_1<\dots<\beta_n\leq b} S^-[\ell(\beta_1),\dots,\ell(\beta_n),\ \ell(b)] + \sup_{b<\gamma_1<\gamma_2<\dots<\gamma_r<c} S^-[\ell(b),\ell(\gamma_1),\dots,\ell(\gamma_r)]$$

$$= S^-_{(a,b]}[\ell] + S^-_{[b,c)}[\ell] \quad .$$

If b is an end point of the interval I , from (7.2.8) , (7.2.9) is trivial.

Lemma 7.2.2. Let $\nu(\alpha)$ and $\nu_0(\alpha)$

(7.2.10) $$\lim_{\alpha\to\infty} \nu(\alpha) = k_\nu \leq \infty \quad ,$$

(7.2.11) $$\nu_0(\alpha) = \begin{cases} \nu(\alpha) & 0\leq\alpha\leq\beta^0 \\ k^* & \beta^0 < \alpha \ , \end{cases} \quad ,$$

and

(7.2.12) $$\nu_0(\alpha) \leq 0 \qquad \text{for } \alpha\leq 0 \quad .$$

Then for

(7.2.13) $$0\leq k\leq k^*\leq k_\nu \quad ,$$

(7.2.14) $$S^-[\nu_0(\alpha)-k] \leq S^-[\nu(\alpha)-k] \quad .$$

Proof: From (7.2.12) and (7.2.13)

$$\nu_0(\alpha)-k\leq 0 \qquad \text{for } \alpha\leq 0 \quad .$$

Thus

(7.2.15) $$S^-_{(-\infty,0]}[\nu_0(\alpha)-k] = 0 \ \leq \ S^-_{(-\infty,0]}[\nu(\alpha)-k] \quad .$$

From (7.2.11)

(7.2.16) $$S^-_{[0,\beta^0]}[\nu_0(\alpha)-k] = S^-_{[0,\beta^0]}[\nu(\alpha)-k] \quad .$$

On the interval $[\beta^0,\infty)$, $\nu_0(\alpha)-k$ takes on only two values $\nu_0(\beta^0)-k$ and k^*-k . If $k=k^*$, then

(7.2.17) $$S^-_{[0,\beta^0]}[\nu_0(\alpha)-k] = S^-[\nu(\beta^0)-k,0] = 0 \leq S^-[\nu(\alpha)-k] \quad .$$

If $k \neq k^*$, then from (7.2.13) , k^*-k and $k_\nu - k$ are both positive. From (7.2.10) there exists $b \in (\beta^0,\infty)$ such that $\nu(b)-k$ is positive. Thus

(7.2.18)
$$S^-_{[\beta^0,\infty)}[\nu_0(\alpha)-k] = S^-[\nu(\beta^0)-k \ , \ k^*-k]$$
$$= S^-[\nu(\beta^0)-k \ , \ \nu(b)-k] \leq S^-_{[\beta^0,\infty)}[\nu(\alpha)-k] \quad .$$

From Lemma 7.2.1, (7.2.15), (7.2.16), (7.2.17) , and (7.2.18)

$$S^-[\nu_0(\alpha)-k] = S^-_{(-\infty,0]}[\nu_0(\alpha)-k] + S^-_{[0,\beta^0]}[\nu_0(\alpha)-k] + S^-_{[\beta^0,\infty]}[\nu_0(\alpha)-k]$$
$$\leq S^-_{(-\infty,0]}[\nu(\alpha)-k] + S^-_{[0,\beta^0]}[\nu(\alpha)-k] + S^-_{[\beta^0,\infty)}[\nu(\alpha)-k]$$
$$= S^-[\nu(\alpha)-k] \quad .$$

The following assumption will be required:

Assumptions (A) : the function $\nu(\alpha)$ is odd, is continuous except for a finite number of discontinuities of the first kind, is bounded on finite intervals, satisfies

(7.2.19) $$\nu(\alpha) \geq 0 \qquad \text{for } 0 \leq \alpha < \infty$$

with strict inequality for some open interval, and

(7.2.20) $$\lim_{\alpha\to\infty} \nu(\alpha) = k_\nu \leq \infty \quad ;$$

further if $k_\nu = \infty$, then assume also that

(7.2.21) $$\bar{N}(\alpha) = \int_{-\infty}^{\infty} |\nu(\alpha+\eta)|n(\eta)d\eta < \infty \qquad \text{for all } \alpha \quad .$$

<u>Lemma</u> 7.2.3. If $\nu(\alpha)$ satisfies (A) with $k_\nu < \infty$, then $\nu(\alpha)$ and $\bar{N}(\alpha)$ are bounded. Further $N(\alpha)$ defined by (7.2.1) is continuous, bounded, odd, and satisfies

(7.2.22) $$N(\alpha) > 0 \qquad \text{for } \alpha > 0 \ ,$$

(7.2.23) $$\lim_{\alpha \to \infty} N(\alpha) = k_\nu \quad ,$$

and

(7.2.24) $$S^+[N(\alpha)-k] \le S^-[\nu(\alpha)-k] \qquad \text{for} \quad -\infty < k < \infty \quad .$$

Proof: Under assumption (A) $\nu(\alpha)$ is piece-wise continuous and bounded on finite intervals; from (7.2.20) for $k_\nu < \infty$ it is bounded and piece-wise continuous on $-\infty < \alpha < \infty$. By assumption $\nu(\alpha)$ is not identically zero a.e. . Let K be a bound for $\nu(\alpha)$, then from (7.2.1) and (7.2.21)

$$|N(\alpha)| \le \bar{N}(\alpha) = \int_{-\infty}^{\infty} |\nu(\alpha+\eta)| n(\eta) d\eta \le K \quad .$$

Thus the integral (7.2.1) exists, and $N(\alpha)$ and $\bar{N}(\alpha)$ are bounded. By assumption (A) $\nu(\alpha)$ is odd. Thus

$$N(-\alpha) = \int_{-\infty}^{\infty} \nu(-\alpha+\eta) n(\eta) d\eta = \int_{-\infty}^{\infty} \nu(-\alpha-\eta) n(-\eta) d\eta$$

$$= - \int_{-\infty}^{\infty} N(\alpha+\eta) n(\eta) d\eta = -N(\alpha) \quad ,$$

and $N(\alpha)$ is odd. For $\alpha > 0$

(7.2.25) $$N(\alpha) = \int_{-\infty}^{\infty} \nu(\alpha+\eta) n(\eta) d\eta = \int_{-\infty}^{\infty} \nu(\eta) n(\eta-\alpha) d\eta$$

$$= \int_{0}^{\infty} \nu(\eta) [n(\eta-\alpha) - n(\eta+\alpha)] d\eta \quad .$$

By assumption $\nu(\eta)$ is non-negative and not identically zero a.e. for $\eta > 0$, and from (7.2.2) for $\alpha > 0$

(7.2.26) $$[n(\eta-\alpha) - n(\eta+\alpha)] > 0 \qquad \text{for} \quad 0 \le \eta < \infty \quad .$$

Thus (7.2.22) follows from (7.2.25) and (7.2.26).

From the bounded convergence theorem

$$\nu(\alpha+\eta) \to k_\nu \quad \text{as} \quad a \to \infty$$

implies that

(7.2.27) $$N(\alpha) = \int_{-\infty}^{\infty} \nu(\alpha+\eta) n(\eta) d\eta \to \int_{-\infty}^{\infty} k_\nu \, n(\eta) d\eta = k_\nu$$

as $\alpha \to \infty$. Similarly, for $|\alpha_0| < \infty$

$$n(\eta-\alpha_m) \to n(\eta-\alpha_0) \quad \text{as} \quad \alpha_m \to \alpha_0$$

implies that

(7.2.28) $$N(\alpha_m) = \int_{-\infty}^{\infty} \nu(\eta)n(\eta-\alpha_m)d\eta \to \int_{-\infty}^{\infty} \nu(\eta)n(\eta-\alpha_0)d\eta$$

$$= N(\alpha_0) \quad \text{as} \quad \alpha_m \to \alpha_0 \quad .$$

In the first case (7.2.27), an integrable bound is provided by

$$|\nu(\alpha+\eta)n(\eta)| \leq Kn(\eta)$$

and in the second case (7.2.28), by

(7.2.29) $$|\nu(\eta)n(\eta-\alpha_m)| \leq K[n(\eta+\bar{\alpha}) + n(\eta-\bar{\alpha}) + \frac{1}{\sqrt{2\pi}\sigma} \chi_{[-\bar{a},\bar{a}]}(\eta)]$$

where

$$\bar{\alpha} = \sup_m |\alpha_m| \quad .$$

Since the α_m converges to a finite limit, $\bar{\alpha}$ is, of course, finite and the right side of (7.2.29) is integrable. Thus $N(\alpha)$ is continuous, bounded, odd, and (7.2.22)-(7.2.23) are satisfied.

For the kernel

(7.2.30) $$K(\alpha,\eta) = n(\eta-\alpha)$$

certainly $K(\alpha,\eta)\nu(\eta)$ is integrable in η for all α, and

(7.2.31) $$N(\alpha)-k = \int_{-\infty}^{\infty} [\nu(\eta)-k]K(\alpha,\eta)d\eta \quad .$$

In Chapter 1, Section 2 of Karlin [10] it is shown that the normal kernel (7.2.30) is strictly totally positive (STP) and hence strictly sign regular ($SS\ R_r$) for all orders r. Thus Theorem 3.1(b) p. 21 of Karlin [10] applies. The result (7.2.24) then follows from (7.2.31).

Lemma 7.2.4. Let $\nu(\alpha)$ satisfy assumptions (A) with $k_\nu = \infty$.

Define

(7.2.32) $$N_m(\alpha) = \int_{-\infty}^{\infty} \nu_m(\alpha+\eta)n(\eta)d\eta$$

(7.2.33) $$N(\alpha) = \int_{-\infty}^{\infty} \nu(\alpha+\eta)n(\eta)d\eta \quad ,$$

where

$$(7.2.34) \qquad \nu_m(\alpha) = \begin{cases} \nu(\alpha) & |\alpha| \leq m \\ m & \alpha > m \\ -m & \alpha < -m \end{cases} .$$

Then

$$(7.2.35) \qquad N_m(\alpha) \to N(\alpha)$$

uniformly on finite intervals, $N(\alpha)$ is continuous, odd, satisfies

$$(7.2.36) \qquad N((\alpha) \geq 0 \quad \text{for} \quad \alpha > 0 \ ,$$

$$(7.2.37) \qquad \lim_{\alpha \to \infty} N(\alpha) = \infty$$

$$(7.2.38) \qquad S^-[N-k] \leq S^-[\nu-k] \quad \text{for} \quad 0 \leq k < \infty \ ,$$

and $\bar{N}(\alpha)$ is bounded on finite intervals.

Proof: From (7.2.32)

$$N_m(\alpha) = \int_{-\infty}^{\infty} \nu_m(\eta) n(\alpha-\eta) d\eta = \int_0^{\infty} \nu_m(\eta) n(\alpha-\eta) d\eta + \int_0^{\infty} \nu_m(-\eta) n(\alpha+\eta) d\eta$$
$$= \int_0^{\infty} \nu_m(\eta) [n(\alpha-\eta) - n(\alpha+\eta)] d\eta .$$

Similarly

$$(7.2.39) \qquad N(\alpha) = \int_0^{\infty} \nu(\eta)[n(\alpha-\eta) - n(\alpha+\eta)] d\eta \ .$$

Thus from (7.2.34)

$$|N(\alpha) - N_m(\alpha)| \leq \int_0^{\infty} |\nu(\alpha) - \nu_m(\eta)| \quad |n(\alpha-\eta) - n(\alpha+\eta)| d\eta$$
$$= \int_m^{\infty} |\nu(\eta) - m| [n(\alpha-\eta) + n(\alpha+\eta)] d\eta$$
$$\leq \int_m^{\infty} [|\nu(\eta)| + m][n(\alpha-\eta) + n(\alpha+\eta)] d\eta \leq \int_m^{\infty} [|\nu(\eta)| + \eta][n(\alpha-\eta) + n(\alpha+\eta)] d\eta \ .$$

For $m > \bar{\alpha}$, $|\alpha| \leq \bar{\alpha}$ and $\eta \geq m$, from (7.2.2) it can easily be shown that

$$(7.2.40) \qquad n(\alpha-\eta) + n(\alpha+\eta) \leq 2n(\eta-\bar{\alpha}) \ .$$

$$(7.2.41) \qquad |N(\alpha) - N_m(\alpha)| \leq 2 \int_m^{\infty} [|\nu(\eta) + \eta] n(\eta-\bar{\alpha}) d\eta$$
$$= 2 \int_{m-\bar{\alpha}}^{\infty} [|\nu(\eta+\bar{\alpha})| + \eta + \bar{\alpha}] n(\eta) d\eta$$

For $\bar{\alpha}$ fixed from (7.2.21) and the properties of the normal density, the right side of (7.2.41) goes to zero as $n \to \infty$. It follows that (7.2.35) holds uniformly for $-\bar{\alpha} \leq \alpha \leq \bar{\alpha}$.

The functions $\nu_m(\alpha)$ satisfy (A) with $k_{\nu_m} = m < \infty$. Thus from Lemma 7.2.3, $N_m(\alpha)$ is continuous, odd

$$N_m(\alpha) > 0 \quad \text{for} \quad 0 < \alpha \quad ,$$

and

$$S^+[N_m(\alpha) - k] \leq S^-[\nu_m(\alpha) - k] \quad \text{for} \quad -\infty < k < \infty \quad . \tag{7.2.42}$$

From (7.2.35) it follows that $N(\alpha)$ is continuous, odd and satisfies (7.2.36). From (7.2.20) with $k_\nu = \infty$, there exists α_M such that

$$\nu(\alpha) \geq M \quad \text{for} \quad \alpha > \alpha_M \quad .$$

From (7.2.39) for $\alpha \geq 0$

$$N(\alpha) \geq \int_{\alpha_M}^{\infty} M[n(\alpha-\eta) - n(\alpha+\eta)]d\eta$$

$$= M[\int_{\alpha_M-\alpha}^{\infty} n(\eta)d\eta - \int_{\alpha_M+\alpha}^{\infty} n(\eta)d\eta] .$$

Thus for M fixed

$$\liminf_{\alpha \to \infty} N(\alpha) \geq M \liminf_{\alpha \to \infty} [\int_{\alpha_M-\alpha}^{\infty} n(\eta)d\eta - \int_{\alpha_M+\alpha}^{\infty} \eta(\eta)d\eta] = M \quad .$$

and (7.2.37) follows.

For $0 \leq \alpha_1 < \alpha_2 < \ldots < \alpha_r$ fixed from Lemma 1.1 p. 217 of Karlin [10], (7.2.35), and the definition (7.2.4)

$$S^-[N(\alpha_1) - k, \ldots, N(\alpha_r) - k] \leq \liminf_{m \to \infty} S^+[N_m(\alpha_1) - k, \ldots, N_m(\alpha_r) - k]$$

$$\leq \liminf_{m \to \infty} S^+[N_m(\alpha) - k] \quad .$$

Thus from (7.2.3) and (7.2.42)

$$S^-[N(\alpha) - k] \leq \liminf_{m \to \infty} S^+[N_m(\alpha) - k] \leq \liminf_{m \to \infty} S^-[\nu_m(\alpha) - k] \quad . \tag{7.2.43}$$

For $0 \leq k$ fixed and $m > k$, $\nu_m(\alpha)$ defined by (7.2.34) satisfies the conditions of Lemma 7.2.2 with $\beta^0 = m$, $k^* = m$, and $k_\nu = \infty$. Thus from that lemma

(7.2.44) $$S^-[\nu_m - k] \leq S^-[\nu - k]$$

for $m > k$. Thus for all $0 \leq k < \infty$, (7.2.38) follows from (7.2.43) and (7.2.44). From (A) $\nu(\alpha)$ is bounded on finite intervals. Let K be the bound of $|\nu(\alpha)|$ on $-\bar{\alpha} \leq \alpha \leq \bar{\alpha}$. Then from (7.2.21), (7.2.40) with $m = \bar{\alpha}$, and (7.2.21)

$$\begin{aligned}
\bar{N}(\alpha) &= \int_0^\infty |\nu(\eta)|[n(\alpha-\eta) - n(\alpha+\eta)]d\eta \\
&\leq K \int_0^{\bar{\alpha}} [n(\alpha-\eta) + n(\alpha+\eta)]d\eta + \int_{\bar{\alpha}}^\infty |\nu(\eta)|[n(\alpha-\eta) + n(\alpha+\eta)]d\eta \\
&\leq K \int_{-\infty}^\infty 2n(\eta)d\eta + 2\int_{\bar{\alpha}}^\infty |\nu(\eta)|n(\eta-\bar{\alpha})d\eta \\
&\leq 2K + 2\int_{-\infty}^\infty |\nu(\eta)|n(\eta-\bar{\alpha})d\eta = 2(K + \bar{N}(\bar{\alpha})] < \infty \quad .
\end{aligned}$$

Thus $\bar{N}(\alpha)$ is bounded on finite intervals.

<u>Lemma</u> 7.2.5. If $\nu(\alpha)$ satisfies assumptions (A) with $k_\nu = \infty$, and $\nu(\alpha)$ is monotone non-decreasing, then $N(\alpha)$ defined by (7.2.1) is continuous, odd, strictly monotone increasing and satisfies (7.2.22) and (7.2.37).

Proof: From Lemma 7.2.4 $N(\alpha)$ is continuous, odd, satisfies (7.2.36) and (7.2.37). It remains to show that $N(\alpha)$ is strictly increasing. Since $\nu(\alpha)$ is monotone non-decereasing and goes to $+\infty$ as $\alpha \to \infty$, it follows that for $\alpha_1 < \alpha_2$

$$\nu(\alpha_2+\eta) - \nu(\alpha_1+\eta) \geq 0 \quad \text{for all} \quad \eta$$

and

$$\nu(\alpha_2+\eta) - \nu(\alpha_1+\eta) \neq 0 \quad \text{a.e.}$$

Thus for $\alpha_2 > \alpha_1$, from (7.2.1)

$$N(\alpha_2) - N(\alpha_1) = \int_{-\infty}^\infty [\nu(\alpha_2+\eta) - \nu(\alpha_1+\eta)]n(\eta)d\eta > 0 \quad .$$

<u>Lemma</u> 7.2.6. Let $\nu(\alpha)$ satisfy assumptions (A) and let

(7.2.45) $$S^-[\nu-k_0] \le 1$$

where $0 \le k_0 \le k_\nu$. If $k_\nu = \infty$, let $\nu(\alpha)$ be monotone non-decreasing. Then either

(7.2.46) $$N(\alpha) - k_0 < 0 \quad \text{for all} \quad \alpha$$

in which case define $\alpha^0 = \infty$, or there exists uniquely $\alpha < \alpha^0$ which satisfies

(7.2.47) $$N(\alpha^0) - k_0 = 0$$

and then

(7.2.48) $$N(\alpha) - k_0 > 0 \quad \text{for} \quad \alpha > \alpha^0$$

(7.2.49) $$N(\alpha) - k_0 < 0 \quad \text{for} \quad \alpha < \alpha^0 \quad .$$

Proof: From Lemmas 7.2.3 and 7.2.4, $N(\alpha)$ is continuous, odd, and satisfies (7.2.36) . If $k_\nu < \infty$, from (7.2.24) of Lemma 7.2.3 and (7.2.45)

$$S^+[N(\alpha) - k_0] \le S^-[\nu-k_0] \le 1 \ .$$

If $k_\nu = \infty$, then from Lemma 7.2.5 $N(\alpha)$ is strictly increasing. This implies that

$$S^+[N(\alpha) - k] = 1 \quad \text{for all} \quad k \quad .$$

Thus in either case

(7.2.50) $$S^+[N(\alpha) - k_0] \le 1 \quad .$$

Suppose there exists $0 \le \alpha_1 < \alpha_2$ such that

$$N(\alpha_1) = N(\alpha_2) = k_0 \quad ,$$

and let $\alpha_3 > \alpha_2$. If $N(\alpha_3) - k_0 \ge 0$, then

$$S^+[N(\alpha_1) - k_0 \ , \ N(\alpha_2) - k_0 \ , \ N(\alpha_3) - k_0]$$
$$= S^+[0^+, 0^-, +] = 2 \quad .$$

The notation 0^+ indicates that the value is zero and that it is assigned the positive sign. It $N(\alpha_3) - k_0 < 0$, then

$$S^+[N(\alpha_1)-k_0 \, , \, N(\alpha_2) - k_0 \, , \, N(\alpha_3) - k_0]$$
$$= S^+[0^-,0^+,-] = 2 \quad .$$

This contradicts (7.2.50) . Thus if there exists a solution to (7.2.47), the solution is unique. Since $N(\alpha)$ is odd.

(7.2.51) $$N(0) = 0 \ .$$

Since $N(\alpha)$ is odd, satisfies (7.2.36), and by assumption $k_0 \geq 0$, it follows that

(7.2.52) $$N(\alpha) - k_0 \leq 0 \quad \text{for} \quad \alpha \leq 0 \ .$$

Suppose there exists α^* such that

(7.2.53) $$N(\alpha^*) - k_0 > 0 \quad .$$

Then from (7.2.52) , $\alpha^* > 0$. From the continuity of $N(\alpha)$, (7.2.52) and (7.2.53) it follows then that there exists α_0 which satisfies (7.2.47) and

$$0 \leq \alpha_0 \leq \alpha^* \quad .$$

Suppose then that $0 \leq \alpha_0$ satisfies (7.2.47) . If there exists α^* such that

$$\alpha^* < \alpha_0$$

and

$$N(\alpha^*) - k_0 \geq 0$$

then for $\alpha^{**} < \alpha^*$ and $\alpha^{**} \leq 0$ from (7.2.52)

$$S^+[N(\alpha^{**}) - k_0 \, , \, N(\alpha^*) - k_0 \, , \, N(\alpha_0) - k_0]$$
$$= S^+[-,+,0^-] = 2$$

contradicting (7.2.50) . Thus (7.2.49) holds. Suppose there exists α^* such that

$$\alpha_0 < \alpha^*$$

and

$$N(\alpha^*) - k_0 \leq 0 \quad ,$$

then

$$S^+[N(0) - k_0 \ , \ N(\alpha^0) - k_0 \ , \ N(\alpha^*) - k_0] = S^+[-,0^+,-] = 2$$

again contradicting (7.2.50), and (7.2.48) follows.

7.3 Absolute value loss: preliminaries

This section is devoted to minimizing the function

$$(7.3.1) \qquad L(\alpha,u) = \int \ell(\alpha + \zeta u + \eta)\, n(\eta,\sigma_1^2)d\eta + \lambda_1|u| + \lambda_2\, \delta(u)$$

in the s dimensional column vector u for each fixed real number α. The function n is the Gaussian density (7.2.2) with mean zero and variance σ_1^2, the absolute value $|u|$ has the usual definition for vectors

$$(7.3.2) \qquad |u| = \sqrt{\sum_{i=1}^{s} u_i^2} \quad ,$$

$\delta(u)$ is defined by

$$(7.3.3) \qquad \delta(u) = \begin{cases} 1 & \text{if } |u| \neq 0 \\ 0 & \text{if } |u| = 0 \end{cases} \quad ,$$

ζ is a given row vector, and the parameters λ_1, λ_2 satisfy

$$7.3.4) \qquad \lambda_1 \geq 0 \ , \ \lambda_2 \geq 0 \ , \ \lambda_1 + \lambda_2 \neq 0 \ .$$

A vector-valued function $\hat{U}(\alpha)$ will be found which satisfies

$$(7.3.5) \qquad L(\alpha,\hat{U}(\alpha)) = \inf_{u \in \mathcal{U}} L(\alpha,u) \qquad -\infty < \alpha < \infty \ .$$

The following assumptions will be required:

<u>Assumptions</u> (B): $\ell(\alpha)$ is non-negative, continuous, satisfies

$$(7.3.6) \qquad \int_{-\infty}^{\infty} \ell(\eta)n(\eta,\sigma^2)d\eta < \infty \qquad \text{for all} \quad \sigma > 0 \ ,$$

and has a derivative at all but a finite number of points

(7.3.7) $$\frac{d\mathscr{L}(\alpha)}{d\alpha} = \nu(\alpha)$$

where $\nu(\alpha)$ satisfies assumptions (A) of the previous section for the variance σ_1^2.

Let

(7.3.8) $$C_{\ell} = \int_{-\infty}^{\infty} \ell(\eta) n(\eta,\sigma_1^2) d\eta \quad .$$

Lemma 7.3.1. Let ℓ satisfy (B) and for $\nu(\alpha)$ and $N(\alpha)$ given by (7.3.7) and (7.2.1) with σ_1^2, and define

(7.3.9) $$W(\beta) = \int_0^{|\beta|} N(\alpha) d\alpha \quad .$$

Then $W(\beta)$ is monotone nondecreasing in $|\beta|$, is differentiable,

(7.3.10) $$\frac{d}{d\beta} W(\beta) = N(\beta) \qquad -\infty < \beta < \infty \quad ,$$

(7.3.11) $$W(\beta) = \int_{-\infty}^{\infty} \ell(\beta+\eta) n(\eta,\sigma_1^2) d\eta - C_{\ell} \qquad -\infty < \beta < \infty \quad ,$$

and if $k_\nu \neq 0$

(7.3.12) $$\lim_{\beta \to \infty} W(\beta) = \infty \quad .$$

Proof: From Lemmas 7.2.3 and 7.2.4, $N(\alpha)$ defined by (7.2.1) is continuous, odd, and non-negative for $\alpha \geq 0$. It follows that $W(\beta)$ is nondecreasing in $|\beta|$ and satisfies (7.3.10) for $\beta \neq 0$. The one-sided derivatives at $\beta = 0$ satisfy

$$\frac{dW}{d\beta^+} = N(\beta+0) \quad , \quad \frac{dW}{d\beta^-} = N(\beta-0) \quad .$$

Since $N(\beta)$ is continuous and $N(0) = 0$, it follows that $N(\beta+0) = N(\beta-0) = 0$ so that (7.3.10) also holds for $\beta = 0$.

Also from Lemmas 7.2.3 and 7.2.4

(7.3.13) $$\lim_{\alpha \to \infty} N(\alpha) = k_\nu \quad .$$

Thus if $k_\nu > 0$, (7.3.12) clearly holds. From Lemmas 7.2.3 and (7.2.4) $\overline{N}(\alpha)$ is bounded on finite intervals. Thus from (7.2.21)

$$\int_0^\beta \int_{-\infty}^{\infty} |\nu(\alpha+\eta)| n(\eta,\sigma_1^2) d\eta d\alpha = \int_0^\beta \bar{N}(\alpha) d\alpha < \infty \quad .$$

Thus the Fubini Theorem can be applied. From (7.3.9), (7.2.1), the Fubini Theorem, (7.3.7), (7.3.6), and (7.3.8) for $\beta \geq 0$

$$\begin{aligned} W(\beta) &= \int_0^\beta N(\alpha) d\alpha = \int_0^\beta [\int_{-\infty}^{\infty} \nu(\alpha+\eta) n(\eta) d\eta] d\alpha \\ &= \int_{-\infty}^{\infty} [\int_0^\beta \nu(\alpha+\eta) d\alpha] n(\eta) d\eta \\ &= \int_{-\infty}^{\infty} [\ell(\beta+\eta) - \ell(\eta)] n(\eta) d\eta \\ &= \int_{-\infty}^{\infty} \ell(\beta+\eta) n(\eta) d\eta - C_\ell \quad . \end{aligned} \tag{7.3.14}$$

For $\beta < 0$, since $\nu(\alpha)$ is odd, from (7.3.7)

$$\begin{aligned} \int_0^{-\beta} \nu(\alpha+\eta) d\alpha &= \int_\beta^0 \nu(-\gamma+\eta) d\gamma \\ &= \int_\beta^0 -\nu(\gamma-\eta) d\gamma = -\ell(-\eta) + \ell(\beta-\eta) \quad . \end{aligned}$$

Thus as in (7.3.14) for $\beta < 0$

$$\begin{aligned} W(\beta) &= \int_0^{-\beta} N(\alpha) d\alpha = \int_{-\infty}^{\infty} [\int_0^{-\beta} \nu(\alpha+\eta) d\alpha] n(\eta) d\eta \\ &= \int_{-\infty}^{\infty} [-\ell(-\eta) + \ell(\beta-\eta)] n(\eta) d\eta = \int_{-\infty}^{\infty} [-\ell(\eta) + \ell(\beta+\eta)] n(\eta) d\eta \\ &= \int_{-\infty}^{\infty} \ell(\beta+\eta) n(\eta) d\eta - C_\ell \quad . \end{aligned}$$

<u>Lemma</u> 7.3.2. Let ℓ satisfy assumption (B) ; if $k_\nu = \infty$, assume further that $\nu(\alpha)$ is monotone non-decreasing. Let $0 \leq k_0 \leq k_\nu$ satisfy

$$S^-[\nu - k_0] \leq 1 \tag{7.3.15}$$

and let α^0 be defined as in Lemma 7.2.6. Then for $\beta \geq \alpha^0$

i)

$$W^*(\beta) = W(\beta) - W(\alpha^0) - k_0(\beta-\alpha^0) = \int_{\alpha^0}^\beta (N(\eta) - k_0) d\eta \tag{7.2.16}$$

is continuous, strictly increasing from $W^*(\alpha^0) = 0$, and if $k_0 < k_\nu$, then

$$\lim_{\beta \to \infty} W^*(\beta) = \infty \quad . \tag{7.3.17}$$

ii) For $0 \leq \lambda_2$ either

$$(7.3.18) \qquad W^*(\beta) < \lambda_2 \quad \text{for all} \quad \beta > \alpha^0$$

in which case let $\beta^0 = \infty$, or there exists uniquely β^0 such that $\alpha^0 \leq \beta^0 < \infty$ and

$$(7.3.19) \qquad W^*(\beta^0) = \lambda_2$$

and then

$$(7.3.20) \qquad \begin{aligned} W^*(\beta) &< \lambda_2 \qquad \alpha^0 \leq \beta < \beta^0 \\ &> \lambda_2 \qquad \beta^0 < \beta \end{aligned} \quad ;$$

if $\alpha^0 = \infty$, let $\beta^0 = \infty$.

iii) If $\lambda_2 > 0$ and $\alpha^0 < \infty$, then $\alpha^0 < \beta^0$.

Proof: Equation (7.3.16) follows from (7.3.9). Since $N(\alpha)$ is continuous. $W^*(\beta)$ is obviously continuous. From (7.2.48) of Lemma 7.2.6 $W^*(\beta)$ is strictly increasing for $\beta > \alpha^0$.

From Lemmas 7.2.3 and 7.2.4, (7.3.13) holds. Thus for $\alpha_0 < \infty$ and $0 \leq k_0 < k_\nu$ (6.3.17) follows from (7.3.16) .

Properties ii) and iii) follow from the fact that $W^*(\beta)$ is continuous and strictly increasing for $\beta \geq \alpha^0$ and $W^*(\alpha^0) = 0$.

Lemma 7.3.3. Let ℓ satisfy assumptions (B) . The $L(\alpha,u)$ given by (7.3.1) satisfies

$$(7.3.21) \qquad L(\alpha,u) = W(\alpha+\zeta u) + \lambda_1|u| + \lambda_2\delta(u) + C_\ell$$

and

$$(7.3.22) \qquad L(-\alpha,-u) = L(\alpha,u) \quad .$$

Proof: Equation (7.3.21) follows from (7.3.1) and (7.3.11) of Lemma 7.3.1; (7.3.22) follows from (7.3.21) since from (7.3.9) $W(\beta)$ is even.

From Lemma 7.3.3 it is sufficient to consider $L(\alpha,u)$ for $0 \leq \alpha$. If $\hat{U}(\alpha)$ satisfies

$$(7.3.23) \qquad L(\alpha,\hat{U}(\alpha)) \leq L(\alpha,u) \qquad \text{for all } u \in \mathcal{U}$$

for $\alpha \geq 0$. Then $\hat{U}(\alpha)$ defined by

$$(7.3.24) \qquad \hat{U}(\alpha) = -\hat{U}(-\alpha)$$

for $\alpha < 0$ satisfies (7.3.23).

It will be assumed hence forth that $\alpha \geq 0$.

<u>Lemma</u> 7.3.4. Let ℓ satisfy (B) . Then if $\zeta \neq 0$

$$(7.3.25) \qquad L\left(\alpha, \frac{\zeta'\rho}{|\zeta|^2}\right) = W(\alpha+\rho) + \frac{\lambda_1|\rho|}{|\zeta|} + \lambda_2\delta(\rho) + C_\ell \leq L(\alpha,u)$$

for all $u \in \mathcal{U}$ such that $\zeta u = \rho$, with strict inequality for $u \neq \frac{\zeta u}{|\zeta|^2}$ provided $\lambda_1 \neq 0$.

Further

$$(7.3.26) \qquad \frac{\partial}{\partial\rho} L\left(\alpha, \frac{\zeta'\rho}{|\zeta|^2}\right) = \begin{cases} N(\alpha+\rho) + \frac{\lambda_1}{|\zeta|} & \text{for } \rho > 0 \\ N(\alpha+\rho) - \frac{\lambda_1}{|\zeta|} & \text{for } \rho < 0 \end{cases} .$$

Proof: Let ρ be a fixed real number, then for u such that

$$\zeta u = \rho$$

from (7.3.21) of Lemma 7.3.3

$$(7.3.27) \qquad L(\alpha,u) = W(\alpha+\rho) + \lambda_1\sqrt{u'u} + \lambda_2\,\delta(u) + C_\ell$$

$$= W(\alpha+\rho) + \lambda_1 \sqrt{\left(\frac{u-\zeta'\rho}{|\zeta|^2}\right)'\left(\frac{u-\zeta'\rho}{|\zeta|^2}\right) + \frac{\rho^2}{|\zeta|^2}} + \lambda_2\delta(u) + C_\ell$$

Thus for

$$(7.3.28) \qquad \hat{u}_\rho = \frac{\zeta'\rho}{|\zeta|^2} ,$$

$$(7.3.29) \qquad L(\alpha,\hat{u}_\rho) \leq L(\alpha,u) \qquad \text{for all } u \ni \zeta u = \rho$$

with strict inequality for $u \neq \hat{u}_\rho$ provided $\lambda_1 \neq 0$. Since $\zeta \neq 0$, $\hat{u}_\rho$ given by (7.3.28) is zero if and only if $\rho = 0$. Thus

$$\delta(\hat{u}_\rho) = \delta(\rho)$$

and (7.3.25) follows from (7.3.27) and (7.3.29) .

Equation (7.3.26) is obtained from (7.3.25) and (7.3.10) of Lemma 7.3.1.

Lemma 7.3.5. Let ℓ satisfy (B) , if $k_\nu = \infty$, let $\nu(\alpha)$ be monotone non-decreasing, let $|\zeta| \neq 0$,

$$\frac{\lambda_1}{|\zeta|} \leq k_\nu \quad , \tag{7.3.30}$$

and

$$S^-[\nu - \frac{\lambda_1}{|\zeta|}] \leq 1 \ . \tag{7.3.31}$$

Then for α^0 defined by Lemma 7.2.6 with $k_0 = \dfrac{\lambda_1}{|\zeta|}$, for $0 \leq \alpha \leq \alpha^0$

$$\frac{\partial L}{\partial \rho}\left(\alpha, \frac{\zeta'\rho}{|\zeta|^2}\right) \begin{array}{ll} > 0 & \text{for } \rho > 0 \\ < 0 & \text{for } \rho < 0 \ , \end{array} \tag{7.3.32}$$

and for $\alpha^0 < \alpha$ and $\rho \neq 0$

$$\frac{\partial L}{\partial \rho}\left(\alpha, \frac{\zeta'\rho}{|\zeta|^2}\right) \begin{array}{ll} > 0 & \alpha^0 - \alpha < \rho \\ < 0 & \alpha^0 - \alpha > \rho \ . \end{array} \tag{7.3.33}$$

Proof: From (7.2.48) and (7.2.49) of Lemma 7.2.6 for $k_0 = \lambda_1/|\zeta|$, and (7.3.26) of Lemma 7.3.4 for $\rho < 0$

$$\frac{\partial L}{\partial \rho}\left(\alpha, \frac{\zeta'\rho}{|\zeta|^2}\right) = N(\alpha+\rho) - \frac{\lambda_1}{|\zeta|} > 0 \quad \text{for} \quad \alpha+\rho > \alpha^0 \tag{7.3.34}$$

$$< 0 \quad \text{for} \quad \alpha+\rho < \alpha^0 \ . \tag{7.3.35}$$

For $0 \leq \alpha \leq \alpha^0$ and $\rho < 0$, $\alpha + \rho < \alpha^0$, thus (7.3.32) in this case follows from (7.3.35). For $\rho > 0$ and $\alpha + \rho > 0$, from (7.3.26) , (7.3.4), and (7.2.22) of Lemmas 7.2.3 and 7.2.5

$$\frac{\partial}{\partial \rho} L\left(\alpha , \frac{\zeta'\rho}{|\zeta|^2}\right) = N(\alpha+\rho) + \frac{\lambda_1}{|\zeta|} > 0 \ ; \ \rho > 0 \ , \ \alpha+\rho > 0 \ . \tag{7.3.36}$$

For $0 \leq \alpha \leq \alpha^0$ and $\rho > 0$, $\alpha + \rho > 0$. Thus (7.3.32) also holds for $\rho > 0$.

For $\alpha^0 < \alpha$ and $0 < \rho$, $0 \leq \alpha^0 < \alpha + \rho$ so (7.3.33) in this case follows from (7.3.36) . For $\alpha^0 < \alpha$ and $\rho < 0$, (7.3.33) follows from (7.3.34) and (7.3.35) .

<u>Lemma</u> 7.3.6. Under the assumptions of Lemma 7.3.5 , for $0 \leq \alpha \leq \alpha^0$

$$L(\alpha,0) = W(\alpha) + C_\ell \leq L(\alpha,u) \tag{7.3.37}$$

for all $u \in \mathcal{U}$ with struct inequality for $u \neq 0$. For $\alpha^0 < \alpha$

$$L\left(\alpha, \frac{\zeta'}{|\zeta|^2}(\alpha^0-\alpha)\right) = W(\alpha_0) + \frac{\lambda_1}{|\zeta|}(\alpha-\alpha_0) + \lambda_2 + C_\ell \leq L(\alpha,u) \tag{7.3.38}$$

for all $u \neq 0$ with strict inequality for $u \neq \frac{\zeta}{|\zeta|^2}(\alpha^0-\alpha)$ and $u \neq 0$ provided $\lambda_1 \neq 0$.

Proof: From (7.3.21) of Lemma 7.3.3, (7.3.25) of Lemma 7.3.4 and (7.3.32) of Lemma 7.3.5 , for $0 \leq \alpha \leq \alpha^0$ and $\rho \neq 0$

$$L(\alpha,0) = W(\alpha) + C_\ell \leq W(\alpha) + \lambda_2 + C_\ell$$

$$= \lim_{|\rho| \downarrow 0} \left(W(\alpha+\rho) + \frac{\lambda_1}{|\zeta|}|\rho| + \lambda_2 + C_\ell \right) = \lim_{|\rho| \downarrow 0} L\left(\alpha, \frac{\zeta'\rho}{|\zeta|^2}\right)$$

$$< L\left(\alpha, \frac{\zeta'\rho}{|\zeta|^2}\right) .$$

Thus (7.3.37) follos from (7.3.25) of Lemma 7.3.4.

For $\alpha^0 < \alpha$ and $\rho \neq 0$, from (7.3.33) of Lemma 7.3.5

$$L\left(\alpha, \frac{\zeta'}{|\zeta|^2}(\alpha^0-\alpha)\right) < L\left(\alpha, \frac{\zeta'\rho}{|\zeta|^2}\right)$$

for $\rho \neq \alpha^0-\alpha$; (7.3.38) follows then from (7.3.25) of Lemma 7.3.4.

<u>Lemma</u> 7.3.7. Under the assumptions of Lemma 7.3.5 α^0 is defined uniquely in Lemma 7.2.6 with $k_0 = \lambda_1/|\zeta|$ and β^0 is defined uniquely in Lemma 7.3.2 ii) . Let $\hat{U}(\alpha)$ satisfy

$$(7.3.39)\qquad \hat{U}(\alpha) = \begin{cases} \dfrac{\zeta'}{|\zeta|^2}(\alpha^0-\alpha) & \text{for } \alpha \geq \beta^0 \\ 0 & \text{for } 0 \leq \alpha < \beta^0 \end{cases}$$

$$\hat{U}(\beta^0) = \frac{\zeta'}{|\zeta|^2}(\alpha^0-\beta^0) \quad \text{or} \quad 0 \quad .$$

Then

$$(7.3.40)\qquad L(\alpha,\hat{U}(\alpha)) = \inf_{u \in \mathcal{U}} L(\alpha,u) \quad .$$

For $\lambda_1 \neq 0$ and $\alpha \neq \beta^0$ there is strict inequality for $u \neq \hat{U}(\alpha)$. For $\lambda_1 \neq 0$ and $\alpha = \beta^0$, there is strict inequality for $u \neq \zeta'(\alpha^0-\beta^0)/|\zeta|^2$ and $u \neq 0$.

Proof: For $0 \leq \alpha \leq \alpha_0$, optimaility of $\hat{U}(\alpha)$ follows from (7.3.37) of Lemma 7.3.6 . In the range $\alpha^0 < \alpha$, from (7.3.38) $L(\alpha,u)$ is minimized by

$$\hat{u} = \frac{\zeta'}{|\zeta|^2}(\alpha^0-\alpha) \quad \text{or} \quad \hat{u} = 0 \quad .$$

From (7.3.37), (7.3.38) of Lemma 7.3.6 and (7.3.16), (7.3.20) of Lemma 7.3.2

$$L(\alpha,0) - L\left(\alpha\ ,\ \frac{\zeta'}{|\zeta|}(\alpha^0-\alpha)\right) = W(\alpha) - W(\alpha_0) - \frac{\lambda_1}{|\zeta|}(\alpha-\alpha_0) - \lambda_2$$

$$= W^*(\alpha) - \lambda_2 < 0 \quad \text{for} \quad \alpha^0 \leq \alpha < \beta^0$$

$$> 0 \quad \text{for} \quad \beta^0 < \alpha \quad .$$

Thus the minimum is attained by $\hat{u} = \frac{\zeta}{|\zeta|}(\alpha^0-\alpha)$ for $\alpha > \beta^0$ and by $\hat{u} = 0$ for $\alpha^0 < \alpha < \beta^0$, as in the definition (7.3.39). For $\alpha = \beta^0$, from (7.3.19) $L(\alpha,0) = L\left(\alpha\ ,\ \frac{\zeta'}{|\zeta|}(\alpha^0-\alpha)\right)$ so either value provides a minimum.

Lemma 7.3.8. Let $\ell(\alpha)$ satisfy the assumptions of Lemma 7.3.5 , let $L_0(\alpha,u)$ be defined by (7.3.1) . Then ℓ_0 defined by

$$(7.3.41)\qquad \ell_0(\alpha) = \inf_{u \in \mathcal{U}} L_0(\alpha,u)$$

satisfies

$$(7.3.42)\qquad \ell_0(0) = C_\ell$$

and assumptions (B) with derivative ν_0 given by

$$(7.3.43)\qquad \nu_0(\alpha) = \begin{cases} \dfrac{\lambda_1}{|\zeta|} & \text{if } \beta^0 < \alpha \\ N(\alpha) & \text{if } 0 \le \alpha \le \beta^0 \end{cases},$$

which satisfies

$$(7.3.44)\qquad k_{\nu_0} = \frac{\lambda_1}{|\zeta|} \quad \text{or} \quad k_\nu \,,\ k_{\nu_0} < \infty \quad ,$$

and

$$(7.3.45)\qquad S^-[\nu_0(\alpha)-k] \le S^-[\nu(\alpha)-k] \quad \text{for} \quad 0 \le k \le \frac{\lambda_1}{|\zeta|} \quad .$$

Proof: From Lemma 7.3.7, (7.3.25) of Lemma 7.3.4, and (7.3.37) of Lemma 7.3.6

$$(7.3.46)\quad \ell_0(\alpha) = L_0(\alpha,\hat{U}(\alpha)) = \begin{cases} W(\alpha^0) + \dfrac{\lambda_1}{|\zeta|}(\alpha-\alpha^0) + \lambda_2 + C_\ell & \text{for } \alpha \ge \beta^0 \\ W(\alpha) + C_\ell & 0 \le \alpha < \beta^0 \end{cases}$$

and from (7.3.22) of Lemma 7.3.3

$$\ell_0(-\alpha) = \ell_0(\alpha) \quad .$$

The initial value (7.3.42) can be obtained from (7.3.46) and (7.3.9) for $\alpha = 0$.

Since $W(\alpha)$ is continuous and non-negative, $\ell_0(\alpha)$ is non-negative and continuous at $\alpha \neq \beta^0$. From the definition (7.3.19) of β^0 , ℓ_0 is also continuous at $\alpha = \beta^0$. From (7.3.46) and (7.3.10) of Lemma 7.3.1, $\ell_0(\alpha)$ has a derivative $\nu_0(\alpha)$ for $\alpha \neq \beta_0$ and ν_0 satisfies (7.3.43) . The function $\nu_0(\alpha)$ is arbitrarily assigned the value of its left one-sided derivative at $\alpha = \beta^0$. Since from Lemmas 7.2.3 and 7.2.4 $N(\alpha)$ is continuous, odd, and satisfies (7.2.36) , ν_0 is piecewise continuous, bounded on finite intervals and satisfies (7.2.19) .

Also from Lemmas 7.2.3 and 7.2.4

$$(7.3.47)\qquad \lim_{\alpha\to\infty} N(\alpha) = k_\nu \ .$$

Thus from (7.3.43)

$$\lim_{\alpha \to \infty} \nu_0(\alpha) = \begin{cases} \frac{\lambda_1}{|\zeta|} & \text{if } \beta^0 < \infty \\ k_\nu & \text{if } \beta^0 = \infty \end{cases} .$$

If $k_\nu = \infty$, then (7.2.46) cannot hold, so $0 \leq \alpha_0 < \infty$ and from (7.3.17) $\beta^0 < \infty$. Thus in this case $k_{\nu_0} = \lambda_1/|\zeta_1|$ and (7.3.44) holds.

If $\beta^0 < \infty$ and $\lambda_1 = 0$, then from assumption (7.3.4) $\lambda_2 \neq 0$ and from Lemma 7.3.2 iii) $0 \leq \alpha^0 < \beta^0$. From (7.3.16) and (7.3.19), if $0 < \beta^0 < \infty$, then $N(\alpha)$ is not identically zero on $0 < \alpha < \beta^0$. Thus from (7.3.43) $\nu_0(\alpha) > 0$ on a nontrivial interval. If $\beta^0 = \infty$ and $k_\nu = 0$, from (7.2.22) of Lemma 7.2.3 $N(\alpha) = \nu_0(\alpha) > 0$ for $0 < \alpha < \infty$. Thus $\nu_0(\alpha)$ satisfies assumption (A) with $k_{\nu_0} < \infty$.

From (7.3.43), (7.3.47), and (7.3.30) the assumptions of Lemma 7.2.2 are satisfied with $\nu(\alpha) = N(\alpha)$, and $k^* = \lambda_1/|\zeta|$. Thus from Lemma 7.2.2, (7.2.5), (7.2.24) of Lemma 7.2.3, and (7.2.38) of Lemma 7.2.4

$$S^-[\nu_0(\alpha)-k] \leq S^-[N(\alpha)-k] \leq S^-[\nu(\alpha)-k]$$

for $0 \leq k \leq \lambda_1/|\zeta|$. Thus (7.3.45) holds.

It remains only to verify (7.3.6) of assumptions (B). If $\beta^0 = \infty$, then from (7.3.46), (7.3.11) of Lemma 7.3.1, properties of the Gaussian density and assumption (7.3.6) for $\ell(\alpha)$, for $\sigma > 0$

$$\int_{-\infty}^{\infty} \ell_0(\eta)n(\eta,\sigma^2)d\eta = \int_{-\infty}^{\infty} [W(\eta) + C_\ell]n(\eta,\sigma^2)d\eta$$

$$= \int_{-\infty}^{\infty} \int_{-\infty}^{\infty} \ell(\eta+\xi)n(\xi,\sigma_1^2)n(\eta,\sigma^2)d\xi\ d\eta$$

$$= \int_{-\infty}^{\infty} \ell(\eta)n(\eta,\sigma^2+\sigma_1^2)d\eta < \infty .$$

For $\beta^0 < \infty$, let K be a bound for $W(\alpha)$ on $|\alpha| \leq \beta^0$. Then from (7.3.46) for $\sigma > 0$

$$\int_{-\infty}^{\infty} \ell_0(\eta) n(\eta,\sigma^2) d\eta \leq (K + C_\ell) \int_{-\beta^0}^{\beta^0} n(\eta,\sigma^2) d\eta$$

$$+ 2 \int_{\beta^0}^{\infty} [W(\alpha_0) + \frac{\lambda_1}{|\zeta|} (\eta - \alpha^0) + \lambda_2 + C_\ell] n(\eta,\sigma^2) d\eta$$

$$\leq \{K + C_\ell + 2[W(\alpha_0) + \lambda_2 + C_\ell]\} \int_{-\infty}^{\infty} n(\eta,\sigma^2) d\eta + \frac{2\lambda_1}{|\zeta|} \int_0^{\infty} \eta \, n(\eta,\sigma^2) d\eta < \infty \quad .$$

7.4 Absolute value loss function: non-decreasing sensitivity

In this section, a linear Gaussian system is considered for which the loss function has one dimensional structure $\mathbf{a}_t$ given by (3.4.17) . The loss function is given by

$$L = c(A_T x_T) + \lambda_1 \sum_{t=0}^{T-1} |u_t| + \lambda_2 \sum_{t=0}^{T-1} \delta(u_t) \tag{7.4.1}$$

where A_T is a $1 \times n$ matrix, the cost function $c(\mathbf{a}_T)$ is non-negative and measurable, the absolute value $|u|$ and the function $\delta(u)$ are given by (7.3.2) and (7.3.3) , and the parameters λ_1 and λ_2 satisfy (7.3.4) of the previous section.

The loss function (7.4.1) has structure $a_t = V_t$ (Definition 3.2.5) where

$$c_T^a(a_T, u_{T-1}) = c(a_T) + \lambda_1 |u_{T-1}| + \lambda_2 \, \delta(u_{T-1}) \tag{7.4.2}$$

$$c_t^a(a_t, u_{t-1}) = \lambda_1 |u_{t-1}| + \lambda_2 \, \delta(u_{t-1}) \qquad t = 0,2,\dots,T-1 \; , \tag{7.4.3}$$

$$a_t = V_t(\mathbf{x}_t, u^{t-1}) = A_T \Phi_{T,t} x_t \quad . \tag{7.4.4}$$

From Theorem 3.4.3 the statistic $\hat{a}_t$ given by (3.4.14) is sufficient for a_t with control transitions $K_t^{\hat{a}}$ and $G_t^{\hat{a},a}$ are one dimensional Gaussian distributions. It will be assumed throughout this section that they have positive variances

$$\bar{\bar{\sigma}}_t^2 = \bar{\bar{\Sigma}}_t > 0 \qquad t = 1,2,\dots,T \tag{7.4.5}$$

$$\bar{\bar{\sigma}}_T^2 = A_T \Pi_T A_T' > 0 \tag{7.4.6}$$

where $\bar{\bar{\Sigma}}_t$ is given by (3.4.20) .

Since the loss function 7.4.1 has the form (6.1.3), it follows from Theorem 6.1.2 that there is a single selection class and there exist universally ϵ-optimal laws $\{\hat{U}^{\hat{a},\epsilon}\}$ which can be computed from (5.4.26)-(5.4.34). Under additional assumptions on the vectors ζ_t (3.4.19) and the cost function $c(a)$, it will be shown that the system has a single strong selection class.

<u>Lemma</u> 7.4.1. Suppose there exist measurable functions $\hat{U}_t(\alpha)$, $L_t(\alpha,u)$, $\ell_t(\alpha)$ which satisfy

(7.4.7) $$\ell_T(\alpha) = c(\alpha)$$

(7.4.8) $$L_t(\alpha,u) = \lambda_1|u| + \lambda_2\delta(u) + \int \ell_{t+1}(\alpha+\zeta_t u+\eta)n_{t+1}(\eta)d\eta$$

(7.4.9) $$\ell_t(\alpha) = \inf_{u\in\mathcal{U}} L_t(\alpha,u)$$

(7.4.10) $$\ell_t(\alpha) = L_t(\alpha,\hat{U}_t(\alpha))$$

$t = T-1, T-2,\dots,0$ where $n_t(\eta) = n(\eta,\bar{\bar{\sigma}}_t^2)$ for $t = 0,\dots,T-1$ and $n_T(\eta) = n(\eta,\bar{\sigma}_T^2 + \bar{\bar{\sigma}}_T^2)$. Then $\hat{U}_t^{\hat{a}}(\alpha) = \hat{U}_t(\alpha)$, $\hat{L}_t^{\hat{a}} = L_t$, $\hat{\ell}_t^{\hat{a}} = \ell_t$ for $t = 0,1,\dots,T-1$ and $\hat{L}_T^{\hat{a}} = \hat{\ell}_T^{\hat{a}} = 0$ satisfies (5.5.21)-(5.5.24) of Definition 5.5.4.

Proof: From Definition 5.2.1, (7.4.2), (3.4.21) of Theorem 3.4.3, and (7.4.6)

(7.4.11) $$c_T^{\hat{a}}(\hat{a}_T,u_{T-1}) = \int_{\mathcal{V}_T} [c(a_T) + \lambda_1|u_{T-1}| + \lambda_2\delta(u_{T-1})]G_T^{\hat{a},a}(\hat{a}_T,da_T)$$
$$= \lambda_1|u_{T-1}| + \lambda_2\delta(u_{T-1}) + \int c(\hat{a}_T+\eta)\bar{n}(\eta)d\eta$$

where $\bar{n}(\eta)$ is the normal density with mean zero and variance $\bar{\sigma}_T^2$. From Definition 5.2.1 and (7.4.3) for $t = 0,1,\dots,T-1$

(7.4.12) $$c_t^{\hat{a}}(\hat{a}_t,u_{t-1}) = \lambda_1|u_{t-1}| + \lambda_2\delta(u_{t-1}) .$$

Let $\hat{U}_t$, L_t, ℓ_t satisfy (7.4.7)-(7.4.10) and let $\hat{U}_t^{\hat{a}}$, $L_t^{\hat{a}}$, and $\ell_t^{\hat{a}}$ be defined as indicated. Clearly (5.5.21) is satisfied. Letting $\bar{\bar{n}}(\eta)$ be the normal density with mean zero and variance $\bar{\bar{\sigma}}_T^2$, from (7.4.8) for

$t = T\text{-}1$, (7.4.7), properties of the normal density, (7.4.11), (5.5.21), (3.4.18), and (7.4.5)

$$\hat{L}^{\hat{a}}_{T-1}(\hat{a}_{T-1},u_{T-1}) = \lambda_1|u_{T-1}| + \lambda_2\delta(u_{T-1}) + \int c(\hat{a}_{T-1}+\zeta_{T-1}u_{T-1}+\eta)n_T(\eta)d\eta$$

$$= \lambda_1|u_{T-1}| + \lambda_2\delta(u_{T-1}) + \int[\int c(\hat{a}_{T-1}+\zeta_{T-1}u_{T-1}+\eta+\eta')\bar{n}(\eta')d\eta']\bar{\bar{n}}(\eta)d\eta$$

$$= \int[\lambda_1|u_{T-1}| + \lambda_2\delta(u_{T-1}) + \int c(\hat{a}_{T-1}+\zeta_{T-1}u_{T-1}+\eta+\eta')\bar{n}(\eta')d\eta']\bar{\bar{n}}(\eta)d\eta$$

$$= \int[c^{\hat{a}}_T(\hat{a}_{T-1}+\zeta_{T-1}u_{T-1}+\eta) + \hat{\ell}^{\hat{a}}_T]\bar{\bar{n}}(\eta)d\eta$$

$$= \int[c^{\hat{a}}_T(\hat{a}_T) + \hat{\ell}^{\hat{a}}_T(\hat{a}_T)]K^{\hat{a}}_T(\hat{a}_{T-1},d\hat{a}_T) \quad .$$

Thus (5.5.22) holds for $t = T\text{-}1$. For $0 \leq t < T\text{-}1$, from (7.4.8), (7.4.12), (3.4.18), and (7.4.5)

$$\hat{L}_t(\hat{a}_t,u_t) = \int[\lambda_1|u_t| + \lambda_2\delta(u_t) + \hat{\ell}^{\hat{a}}_{t+1}(\hat{a}_t+\zeta_t u_t+\eta)]n_{t+1}(\eta)d\eta$$

$$= \int[c^{\hat{a}}_{t+1}(\hat{a}_t+\zeta_t u_t+\eta,u_t) + \hat{\ell}^{\hat{a}}_{t+1}(\hat{a}_t+\zeta_t u_t+\eta)]n_{t+1}(\eta)d\eta$$

$$= \int[c^{\hat{a}}_{t+1}(\hat{a}_{t+1},u_t) + \hat{\ell}^{\hat{a}}_{t+1}(\hat{a}_{t+1})]K^{\hat{a}}_t(\hat{a}_t,d\hat{a}_{t+1}) \quad .$$

Thus (5.5.22) holds for $0 \leq t < T\text{-}1$. Equations (5.5.23) and (5.5.24) follow trivially from (7.4.9) and (7.4.10).

Lemma 7.4.2. Let $c(\alpha)$ have the following properties

i) $\int_{-\infty}^{\infty} c(\eta)n(\eta,\sigma^2)d\eta < \infty$ for all $\sigma > 0$;

ii) $c(\alpha)$ is non-negative, continuous, and has a derivative at all but a finite number of points

$$\frac{d}{d\alpha}c(\alpha) = \nu(\alpha) \quad ;$$

iii) $\nu(\alpha)$ is odd, piece-wise continuous, has a finite number of discontinuities, is bounded on finite intervals and

$$0 \leq \nu(\alpha) \qquad \text{for } 0 \leq \alpha$$

with strict inequality for some open interval;

iv) $\lim_{\alpha \to \infty} \nu(\alpha) = k_T \leq \infty$, $k_T \neq 0$;

v) if $k_T = \infty$, then

$$\int_{-\infty}^{\infty} |\nu(\alpha+\eta)| n_T(\eta) d\eta < \infty \quad \text{for all } \alpha ;$$

and

vi) $S^-[\nu(\alpha)-k] \leq 1$ for all $0 \leq k \leq k_T$.

Assume further that the vectors ζ_t satisfy

$$|\zeta_0| \geq |\zeta_1| \geq \dots \geq |\zeta_{T-1}| > 0 \tag{7.4.13}$$

and

$$|\zeta_{T-1}| \geq \frac{\lambda_1}{k_T} \quad . \tag{7.4.14}$$

Then there exist uniquely $0 \leq \alpha_t^0 \leq \beta_t^0 \leq \infty$, $t = 0,1,\dots,T-1$ determined by a backward iteration as follows: let

$$\ell_T(\alpha) = c(\alpha) \quad ; \tag{7.4.15}$$

$$\nu_{t+1}(\alpha) = \frac{d}{d\alpha} \ell_{t+1}(\alpha) \quad ; \tag{7.4.16}$$

if

$$\int_{-\infty}^{\infty} \nu_{t+1}(\alpha+\eta) n_{t+1}(\eta) d\eta < \frac{\lambda_1}{|\zeta_t|} \quad \text{for all } 0 \leq \alpha < \infty \quad , \tag{7.4.17}$$

let $\alpha_t^0 = \beta_t^0 = \infty$, otherwise α_t^0 is defined by

$$\int_{-\infty}^{\infty} \nu_{t+1}(\alpha_t^0 + \eta) n_{t+1}(\eta) d\eta = \frac{\lambda_1}{|\zeta_t|} \quad ; \tag{7.4.18}$$

for $\alpha_t^0 < \infty$, if

$$\int_{-\infty}^{\infty} [\ell_{t+1}(\beta+\eta) - \ell_{t+1}(\alpha_t^0+\eta)] n_{t+1}(\eta) d\eta - \frac{\lambda_1}{|\zeta_t|} (\beta - \alpha_0^t) < \lambda_2 \tag{7.4.19}$$

$$\text{for all } \alpha_t^0 \leq \beta < \infty \quad ,$$

then let $\beta_t^0 = \infty$. Otherwise β_t^0 is defined by

$$\int_{-\infty}^{\infty} [\ell_{t+1}(\beta_t^0+\eta) - \ell_{t+1}(\alpha_t^0+\eta)] n_{t+1}(\eta) d\eta - \frac{\lambda_1}{|\zeta_t|} (\beta - \alpha_0^t) = \lambda_2 \quad ; \tag{7.4.20}$$

ν_t is defined by

$$(7.4.21)\qquad \nu_t(\alpha) = \begin{cases} \dfrac{\lambda_1}{|\zeta_t|} & \text{for} \quad \beta_t^0 < \alpha \\ \displaystyle\int_{-\infty}^{\infty} \nu_{t+1}(\eta+\alpha) n_{t+1}(\eta) d\eta & \text{for} \quad 0 \leq \alpha \leq \beta_t^0 \end{cases} \quad ;$$

$$(7.4.22)\qquad \nu_t(-\alpha) = -\nu_t(\alpha) \quad ;$$

and $\ell_t(T)$ is determined from (7.4.16) and

$$(7.4.23)\qquad \ell_t(0) = \int_{-\infty}^{\infty} \ell_{t+1}(\eta) n_{t+1}(\eta) d\eta \; .$$

Further, if $\hat{U}_t(\alpha)$ satisfies

$$(7.2.24)\qquad \hat{U}_t(\alpha) = \begin{cases} \dfrac{\zeta_t'}{|\zeta|^2}(\alpha_t^0 - \alpha) & \text{for} \quad \beta_t^0 < \alpha \\ 0 & |\alpha| < \beta_t^0 \\ \dfrac{\zeta_t'}{|\zeta_t|^2}(-\alpha_t^0 - \alpha) & \alpha < -\beta_t^0 \end{cases}$$

$$(7.4.25)\qquad \hat{U}_t(\beta_t^0) = \frac{\zeta_t'}{|\zeta_t|^2}(\alpha_t^0 - \beta_t^0) \quad \text{or} \quad 0$$

$$(7.4.26)\qquad \hat{U}_t(-\beta_t^0) = \frac{\zeta_t'}{|\zeta_t|^2}(-\alpha_t^0 + \beta_t^0) \quad \text{or} \quad 0 \; ,$$

then ℓ_t and $\hat{U}_t$ defined above and L_t defined by (7.4.8) satisfy (7.4.7)-(7.4.10) .

Proof: It will be shown by backward induction from $t = T-1$ that

I-i) α_t^0 and β_t^0 are uniquely defined by (7.4.17)-(7.4.20) ;

I-ii) ℓ_t and ν_t defined by (7.4.21)-(7.4.23) and (7.4.16) for t satisfies assumption (B) with $k_t = k_{\nu_t} < \infty$

I-iii) $\dfrac{\lambda_1}{|\zeta_t|} \leq k_t$

I-iv) $S^-[\nu_t - k] \leq 1 \qquad$ for $\quad 0 \leq k \leq \dfrac{\lambda_1}{|\zeta_t|}$

I-v) ℓ_t defined by (7.4.21)-(7.4.23), (7.4.16) and $\hat{U}_t(\alpha)$ given by (7.4.24)-(7.4.26) satisfy (7.4.9) and (7.4.10), where L_t is computed by (7.4.8) from ℓ_{t+1} .

First, it will be shown that the assumptions i)-vi) on $c(\alpha)$ assure that $\ell_T(\alpha) = c(\alpha)$, $\nu_T(\alpha) = \nu(\alpha)$ satisfy the assumptions of Lemma 7.3.5. Assumptions i)-iv) imply that (B) is satisfied with $\sigma_1^2 = \bar{\sigma}_T^2 + \bar{\bar{\sigma}}_T^2$. If $k_\nu = k_T = \infty$, then assumption vi) together with the fact that $\nu(\alpha)$ is odd implies that $\nu(\alpha)$ is monotone non-decreasing. Assumptions (7.3.30) and (7.3.31) for $\zeta = \zeta_{T-1}$ follow from (7.4.14) and vi) . Thus Lemmas 7.3.5-7.3.8 hold. From Lemma 7.3.7 α^0_{T-1} and β^0_{T-1} are defined uniquely by (7.4.17)-(7.4.20) and $\hat{U}_{T-1}$ given by (7.4.24)-(7.4.26) satisfies (7.4.10) where L_{T-1} and ℓ_{T-1} are defined by (7.4.8) and (7.4.9) . The symmetric results for α negative are obtained from (7.3.24) and (7.3.22). From Lemma 7.3.8 ℓ_{T-1} defined by (7.4.9) satisfies assumption (B) with ν_{T-1} given by (7.4.21) and

$$(7.4.27) \qquad k_{T-1} = k_{\nu_0} = \frac{\lambda_1}{|\zeta_{T-1}|} \quad \text{or } k_T \text{ and } k_{T-1} < \infty \quad .$$

Equations (7.4.22) and (7.4.16) follow from assumption (B) , and (7.4.23) from (7.3.42) of Lemma 7.3.8. It has now been established that α^0_{T-1} , β^0_{T-1} , ℓ_{T-1} , ν_{T-1} , and $\hat{U}_{T-1}$ satisfy I-i) , I-ii) , and I-v) . From (7.3.45) of Lemma 7.3.8 , vi) , and (7.4.14)

$$S^-[\nu_{T-1}(\alpha)-k] \leq S^-[\nu_T(\alpha)-k] \leq 1 \quad \text{for} \quad 0 \leq k \leq \frac{\lambda_1}{|\zeta_{T-1}|} \quad .$$

Thus I-iv) holds for ν_{T-1} . I-iii) follows from (7.4.27) and (7.4.14) . Thus the induction hypothesis holds for $t = T-1$. Assume then that I-i)-I-v) holds for ℓ_{t+1} , ν_{t+1} , k_{t+1} , ζ_{t+1} . From I-iii) of the induction hypothesis and (7.4.13)

$$(7.4.28) \qquad k_{t+1} \geq \frac{\lambda_1}{|\zeta_{t+1}|} \geq \frac{\lambda_1}{|\zeta_t|} \quad .$$

Thus from I-ii) and I-iv) of the induction hypothesis the assumptions of Lemma 7.3.5 are satsfied for $\ell = \ell_{t+1}$, $\nu = \nu_{t+1}$, $\zeta = \zeta_t$, and $k_\nu = k_{t+1}$. As before Lemmas 7.3.7 and 7.3.8 apply. As above from these lemmas it follows that I-i) , I-ii), and I-v) hold for ℓ_t , ν_t and ζ_t . Also,

$$. \; k_t = \frac{\lambda_1}{|\zeta_t|} \quad \text{or} \quad k_{t+1} \quad .$$

I-iii) then follows from (7.4.28) . From (7.3.45) of Lemma 7.3.8 and I-iv) of the induction hypothesis

$$S^-[\nu_t - k] \leq S^-[\nu_{t+1} - k] \leq 1$$

for $0 \leq k \leq \frac{\lambda_1}{\zeta_t}$ and $0 \leq k \leq \frac{\lambda_1}{\zeta_{t+1}}$. I-iv) then follows from (7.4.28).

Theorem 7.4.1. Under the assumptions of Lemma 7.4.2 the system has a single strong selection class in $\hat{\alpha}_t$ and $\hat{U}^{\hat{\alpha}_t}(\alpha) = \hat{U}_t(\alpha)$ given by (7.4.24)-(7.4.26) is a universal optimal law provided the laws $\{u^{t-1}\} \vee \{\hat{U}\}$ are admissible for all admissible $\{u\}$.

Proof: This follows from Lemmas 7.4.1 and 7.4.2. Definition 5.5.4, and Theorem 5.5.5 .

Theorem 7.4.2. Let the assumptions of Lemma 7.4.2 hold and in addition let

(7.4.29) $\qquad \lambda_1 > 0$.

Then

i)

(7.4.30) $\qquad 0 < \alpha_t^0$

and the optimal law $\{\hat{U}\}$ satisfying (7.4.7)-(7.4.10) is given uniquely by (7.4.24)-(7.4.26) .

ii) If

(7.4.31) $\qquad \lambda_2 = 0$,

then

(7.4.32) $\qquad \alpha_t^0 = \beta_t^0$.

iii) If

$$\frac{\lambda_1}{|\zeta_{T-1}|} < k_T \quad \text{and} \quad |\zeta_t| < |\zeta_{t-1}| \quad , \quad t = 1,2,\ldots,T-1 \quad , \tag{7.4.33}$$

then

$$0 < \alpha_t^0 \leq \beta_t^0 < \infty \quad . \tag{7.4.34}$$

iv) If in addition to (7.4.34)

$$\lambda_2 > 0 \; , \tag{7.4.35}$$

then

$$0 < \alpha_t^0 < \beta_t^0 < \infty \quad . \tag{7.4.36}$$

Proof: i) Since ν_{t+1} is odd ,

$$\int_{-\infty}^{\infty} \nu_{t+1}(\eta) n_{t+1}(\eta) d\eta = 0 \quad . \tag{7.4.37}$$

Thus if α_t^0 satisfies (7.4.18) for $\lambda_1 > 0$, it follows that $\alpha_t^0 \neq 0$ and hence (5.4.30) holds. The uniqueness of the optimum follows from Lemma 7.3.7.

ii) For $\lambda_2 = 0$, (7.4.32) clearly provides a solution to (7.4.20) .

iii) Under the assumptions (7.4.29) and (7.4.33) it will be shown by induction that

$$k_t = \frac{\lambda_1}{|\zeta_t|} < \infty \tag{7.4.38}$$

and (7.4.34) holds for $t = T-1, T-2, \ldots$. From (7.2.23) of Lemma 7.2.3, (7.2.37) of Lemma 7.2.4, and (7.4.33)

$$\lim_{\alpha \to \infty} N_T(\alpha) = k_T > \frac{\lambda_1}{|\zeta_{T-1}|} \quad .$$

Thus in Lemma 7.2.6 (7.2.46) cannot hold for $k_0 = \lambda_1/|\zeta_{T-1}|$, so there exists $0 < \alpha_{T-1}^0 < \infty$ which satisfies (7.2.47) . From Lemma 7.3.2 i) , since

$$k_0 = \frac{\lambda_1}{|\zeta_{T-1}|} < k_\nu = k_T \quad ,$$

$$\lim_{\beta \to \infty} W_T^*(\beta) = \infty \quad .$$

Thus (7.3.18) cannot hold and there exist $\alpha^0_{T-1} \leq \beta^0_{T-1} < \infty$ which satisfies (7.3.19) . Since $\beta^0_{T-1} < \infty$, from (7.4.21)

$$\nu_{T-1}(\alpha) = \frac{\lambda_1}{|\zeta_{T-1}|} \qquad \text{for} \quad \alpha > \beta^0_{T-1}$$

and (7.4.38) follows for t = T-1 . Assume then that (7.4.34) and (7.4.38) hold for t+1 where $t+1 \leq T-1$. From (7.2.23) of Lemma 7.2.3 and the induction hypothesis (7.4.38)

$$\lim_{\alpha \to \infty} N_{t+1}(\alpha) = k_{t+1} = \frac{\lambda_1}{|\zeta_{t+1}|}$$

From (7.4.29) and (7.4.33)

$$(7.4.39) \qquad \frac{\lambda_1}{|\zeta_t|} < \frac{\lambda_1}{|\zeta_{t+1}|} \quad .$$

Thus (7.2.46) of Lemma 7.2.6 cannot hold for $k_0 = \lambda_1/|\zeta_t|$, and there must exist α^0_t which satisfies $0 < \alpha^0_t < \infty$ and (7.2.47) . From (7.4.39) and the induction hypothesis (7.4.38)

$$k_0 = \frac{\lambda_1}{|\zeta_t|} < \frac{\lambda_1}{|\zeta_{t+1}|} = k_{t+1} = k_\nu \quad .$$

Thus from Lemma 7.3.2 i)

$$\lim_{\beta \to \infty} W^*_{t+1}(\beta) = \infty \quad .$$

Thus (7.3.18) cannot hold and there exists $\alpha^0_t \leq \beta^0_t < \infty$, which satisfies (7.3.19) . Since $\beta^0_t < \infty$, from (7.4.21)

$$\nu_t(\alpha) = \frac{\lambda_1}{|\zeta_t|} \qquad \text{for} \quad \alpha > \beta^0_t$$

and (7.3.38) follows for t .

iv) Under the additional assumption (7.4.35), (7.4.36) follows from (7.3.16) and (7.3.19) of Lemma 7.3.2.

It should be noted that i) of Theorem 7.4.2 asserts uniqueness of $\{\hat{U}\}$ satisfying (7.4.7)-(7.4.10) and not uniqueness of the universal optimal law $\{\hat{U}^{\hat{a}}\}$ which satisfies the definition of the single strong selection class. From Definition 5.5.4 it is clear that only a.s. uniqueness is possible. Under the assumptions of Theorem 7.4.2 it can be shown that $\hat{U}_t^{\hat{a}}(\hat{a}_t)$ satisfies the Definition 5.5.4 if and only if it is equal to (7.4.24) a.s. $P_{\{u\}}$ for all control laws. This will not be proved rigorously as it is of little interest. Necessary and sufficient conditions for the optimality of a control law $\{\hat{u}\}$ are provided by ii) and iii) of Theorem 5.5.5 where the functions $\hat{L}_t^{\hat{a}} = L_t$ and $\boldsymbol{\ell}_t^{\hat{a}} = \boldsymbol{\ell}_t$ are given by Lemma 7.4.2.

Theorem 7.4.3. Let the assumptions of Lemma 7.4.2 be satisfied, and let

(7.4.40) $$\lambda_1 = 0$$

(7.4.41) $$\lambda_2 > 0 .$$

Then α_t^0 and β_t^0 computed by (7.4.15)-(7.4.23) satisfy

(7.4.42) $$\alpha_t^0 = 0 < \beta_t^0 \leq \infty ,$$

and any measurable control law $\{\hat{U}\}$ which satisfies

(7.4.43) $$\zeta_t \hat{U}_t(\alpha) = \begin{cases} -\alpha & |\alpha| > \beta_t^0 \\ 0 & |\alpha| < \beta_t^0 \end{cases}$$

(7.4.44) $$\zeta_t \hat{U}_t(\beta_t^0) = -\beta_t^0 \quad \text{or} \quad 0$$

(7.4.45) $$\zeta_t \hat{U}_t(-\beta_t^0) = \beta_t^0 \quad \text{or} \quad 0$$

is a universal optimal law provided $\{u^{t-1}\} \vee \{\hat{U}\}$ are admissible for all admissible $\{u\}$.

Proof: From (7.4.37) and (7.4.40) , $\alpha_t^0 = 0$ satisfies (7.4.18) . From (7.4.8) and (7.4.40)

(7.4.46) $$L_t(\alpha,u) = \int \boldsymbol{\ell}_{t+1}(\alpha + \zeta_t u + \eta) n_{t+1}(\eta) d\eta + \lambda_2 \delta_{(u)} .$$

Let $\hat{U}_t(\alpha)$ satisfy (7.4.43)-(7.4.45) and let $\hat{\hat{U}}_t(\alpha)$ satisfy (7.4.24)-(7.4.26) where the same options are observed. Then

$$\zeta_t \hat{\hat{U}}_t(\alpha) = \zeta_t \hat{U}_t(\alpha)$$

and

$$|\hat{U}_t(\alpha)| = 0 \quad \text{iff} \quad |\hat{\hat{U}}_t(\alpha)| = 0 \quad .$$

Thus from (7.4.46)

$$L_t(\alpha, \hat{\hat{U}}_t(\alpha)) = \hat{L}_t(\alpha, \hat{U}_t(\alpha))$$

and the result follows from Lemmas 7.4.1 and 7.4.2, Definition 5.5.4 and Theorem 5.5.5 .

Appendix

A.1 Stochastic kernels

It will be assumed throughout this section that $(\mathcal{X},\mathcal{B}_{\mathcal{X}})$, $(\mathcal{Y},\mathcal{B}_{\mathcal{Y}})$, $(Z,\mathcal{B}_{Z})$, and $(\mathcal{W},\mathcal{B}_{\mathcal{W}})$ are measurable spaces which bear no relation to the spaces of the main body of the monograph. With the exception of Lemma A.1.9 , all the results of this section are available in the literature. They are collected here for easy reference.

Definition A.1.1 . $Q(x,B)$ is a stochastic kernel on $(\mathcal{X},\mathcal{Y})$ provided

i) for each $x \in \mathcal{X}$ fixed $Q(x,B)$ is a probability measure on $(\mathcal{Y},\mathcal{B}_{\mathcal{Y}})$ and

ii) for each $B \in \mathcal{B}_{\mathcal{Y}}$ fixed $Q(x,B)$ is a $\mathcal{B}_{\mathcal{X}}$ -measurable function in x . A semi-colon will be used to separate the arguments if the first argument is an n -tuple . For example $H(x,y;C)$.

Lemma A.1.1 . Let $Q(x,B)$ be a stochastic kernel on $(\mathcal{X},\mathcal{Y})$ and let $f(x,y)$ be non-negative and $\mathcal{B}_{\mathcal{X}} \times \mathcal{B}_{\mathcal{Y}}$ -measurable ($+\infty$ is admitted). Then

$$h(x) = \int_{\mathcal{Y}} f(x,y)Q(x,dy) \tag{A.1.1}$$

is non-negative and $\mathcal{B}_{\mathcal{X}}$ -measurable.

Proof: The result will be shown first for the characteristic (identity) function

$$f(x,y) = \chi_{A\times B}(x,y) \qquad A \in \mathcal{B}_{\mathcal{X}} , B \in \mathcal{B}_{\mathcal{Y}} .$$

Then

$$h(x) = \int_{\mathcal{Y}} \chi_{A\times B}(x,y)Q(x,dy) = \chi_A(x)Q(x,B)$$

is clearly measurable. For $f = \chi_C$ where C is in the field of the product sets (finite disjoint unions) the result follows from the linearity of the integral. From the monotone convergence theorem (Loève [12] p. 124) it can be shown that the class of sets C for which the result is true is a monotone class and hence that the result holds for $f = \chi_C$ and

$C \in \mathcal{B}_{\mathcal{Y}} \times \mathcal{B}_Z$. Next, the result is extended to f a simple function, and then again using the monotone convergence theorem it is shown to hold for $f(x,y)$ non-negative and jointly measureable.

<u>Lemma</u> A.1.2 i) Let $Q(x,B)$ be a stochastic kernel on $(\mathcal{X},\mathcal{Y})$ and let $H(x,y;C)$ be jointly measureable in (x,y) for all C and a measure (not necessarily σ-finite) in $C \in \mathcal{B}_Z$ for all (x,y) . Then

$$(A.1.2) \qquad J(x,C) = \int_{\mathcal{Y}} H(x,y;C)Q(x,dy)$$

is measurable in x for all C and a measure in C for all x .

ii) If $H(x,y;C)$ is a stochastic kernel on $(\mathcal{X} \times \mathcal{Y},Z)$ then $J(x,C)$ is a stochastic kernel on $(\mathcal{X},Z)$.

Proof: i) From Lemma A.1.1 $J(x,C)$ is $\mathcal{B}_{\mathcal{X}}$-measurable in x for all C . For x fixed, from the finite additiveity of $H(x,y;C)$, $J(x,C)$ is finitely additive in C . For $C_n \uparrow C_0$, $C_n \in \mathcal{B}_Z$, $C_0 \in \mathcal{B}_Z$

$$H(x,y;C_n) \uparrow H(x,y;C_0)$$

for all (x,y) since H is a measure in C . Thus from the monotone convergence theorem $J(x,C)$ is continuous from below in C . It follows then from A. p. 84 of Loève [12] that $J(x,C)$ is a measure in C for all x .

ii) If $H(x,y,C)$ is a stochastic kernel, then $H(x,y,z) = 1$ for all (x,y) and from (A.1.2) $J(x,Z) = 1$ for all x .

<u>Lemma</u> A.1.3 . Let $Q(x,A)$ and $H(x,y;B)$ be stochastic kernels on $(\mathcal{X},\mathcal{Y})$ and $(\mathcal{X} \times \mathcal{Y},Z)$ respectively. Then for $A \in \mathcal{B}_{\mathcal{Y}}$ and $B \in \mathcal{B}_Z$

$$(A.1.3) \qquad J(x,A \times B) = \int_A H(x,y;B)Q(x,dy)$$

uniquely determines a stochastic kernel on $(\mathcal{X},\mathcal{Y} \times Z)$.

Proof: It can easily be shown that $\chi_C(y,z)$ is a stochastic kernel on $((\mathcal{X} \times \mathcal{Y}) \times Z , \mathcal{Y} \times Z)$. Thus from Lemma A.1.2 ii)

$$(A.1.4) \qquad H^*(x,y;C) = \int \chi_C(y,z)H(x,y,dz)$$

is a stochastic kernel on $(\mathfrak{X} \times \mathfrak{Y}, \mathfrak{Y} \times Z)$. Again from Lemma A.1.2 ii)

$$J^*(x,C) = \int_{\mathfrak{Y}} H^*(x,y;C)\,Q(x,dy) \tag{A.1.5}$$

is a stochastic kernel on $(\mathfrak{X}, \mathfrak{Y} \times Z)$. For $C = A \times B$, $A \in \mathfrak{B}_{\mathfrak{Y}}$, $B \in \mathfrak{B}_Z$ from (A.1.4) , (A.1.5) , and (A.1.3)

$$H^*(x,y;A \times B) = \int_Z \chi_A(y)\chi_B(z)H(x,y;dz)$$

$$= \chi_A(y)H(x,y;B)$$

$$J^*(x,y;A \times B) = \int_{\mathfrak{Y}} \chi_A(y)H(x,y;B)Q(x,dy)$$

$$= \int_A H(x,y;B)Q(x,dy) = J(x,y;A \times B) .$$

Since for each (x,y) $J(x,y;A \times B)$ is a probability measure on the product sets, it can be extended uniquely to the measure $J^*(x,y;C)$ on $C \in \mathfrak{B}_{\mathfrak{Y}} \times \mathfrak{B}_Z$.

Lemma A.1.4 . Let $Q(x,C)$ be a stochastic kernel on $(\mathfrak{X}, \mathfrak{W})$ and let

$$\psi = \mathfrak{X} \times \mathfrak{W} \to \mathfrak{Y} \tag{A.1.6}$$

be a measurable transformation, then

$$H(x,B) = Q(x;\{w \mid \psi(x,w) \in B\}) \tag{A.1.7}$$

is a stochastic kernel on $(\mathfrak{X}, \mathfrak{Y})$.

Proof: Since $\chi_B(\psi(x,w)$ is a stochastic kernel on $(\mathfrak{X} \times \mathfrak{W}, \mathfrak{Y})$, from Lemma A.1.2 ii)

$$H(x,B) = \int \chi_B(\psi(x,w))Q(x,dw)$$

is a stochastic kernel on $(\mathfrak{X}, \mathfrak{Y})$.

Lemma A.1.5 . Let

$$\begin{aligned} &X:\mathfrak{W} \to \mathfrak{X} \\ &Y:\mathfrak{W} \to \mathfrak{Y} \end{aligned} \tag{A.1.8}$$

be random variables on the probability space $(\mathfrak{W}, \mathfrak{B}_{\mathfrak{W}}, P_{\mathfrak{W}})$, let $P_{\mathfrak{X}}$ be the probability on $(\mathfrak{X},\mathfrak{B}_{\mathfrak{X}})$ induced by $X(w)$, let $Q(x,B)$ be a stochastic kernel on $(\mathfrak{X},\mathfrak{Y})$ which satisfies

(A.1.9) $$Q(x,B) = P_{\mathcal{W}}[Y \in B | X = x] \qquad \text{a.s. } P_{\mathfrak{X}}$$

for all $B \in \mathfrak{B}_{\mathfrak{Y}}$, and let $f(x,y)$ be non-negative and $\mathfrak{B}_{\mathfrak{X}} \times \mathfrak{B}_{\mathfrak{Y}}$ -measurable. Then $h(x)$ defined by (A.1.1) satisfies

(A.1.10) $$h(x) = E_{\mathcal{W}}[f(X,Y) | X = x] \qquad \text{a.s. } P_{\mathfrak{X}} \quad .$$

(Definition of the conditial probabilities and expectations follows Lehman [11] 2.4.)

Proof: The proof follows that of Lemma A.1.1 very closely. For $f = \chi_{A \times B}$ from the properties of conditional expectation, (A.1.9) and (A.1.1)

$$E_{\mathcal{W}}[\chi_A(X)\chi_B(Y) | X = x] = \chi_A(x)E_{\mathcal{W}}[\chi_B(Y) | X = x]$$

$$= \chi_A(x)Q(x,B) = \int \chi_A(x)\chi_B(y)Q(x,dy) = h(x) \qquad \text{a.s. } P_{\mathfrak{X}} \quad .$$

From the linearity of the conditional expectation the result (A.1.10) holds for $f = \chi_C$ where C is in the field over the product sets. The class of the sets C for which (A.1.10) holds with $f = \chi_C$ is shown to be monotone using the monotone convergence theorem for conditional expectation (Loève [12] p.348). Then from the linearity of the conditional expectation and the monotone convergence theorem of conditional expectations again, the result (A.1.10) is extended first to simple functions and finally to non-negative measurable f.

Lemma A.1.6 . Under the assumptions of Lemma A.1.5

(A.1.11) $$P_{\mathcal{W}}[X \in A \ , \ Y \in B] = \int_A Q(x,B)P_{\mathfrak{X}}(dx)$$

for all $A \in \mathfrak{B}_{\mathfrak{X}}$ and $B \in \mathfrak{B}_{\mathfrak{Y}}$, and

(A.1.12) $$E_{\mathcal{W}}[f(X,Y)] = \int_{\mathfrak{X}} [\int_{\mathfrak{Y}} f(x,y)Q(x,dy)]P_{\mathfrak{X}}(dx) \ .$$

Proof: The result (A.1.11) follows immediately from the definition of the conditional probability (A.1.9). From the defining property of the conditional expectation and (A.1.1)

$$E_{\mathcal{W}}[f(X,Y)] = \int h(x)P_{\mathfrak{X}}(dx)$$

$$= \int [\int f(x,y)Q(x,dy)]P_{\mathfrak{X}}(dx) \quad .$$

<u>Lemma</u> A.1.7 . Let

$$\text{(A.1.13)} \qquad \begin{aligned} X &: \mathcal{W} \to \mathfrak{X} \\ Y &: \mathcal{W} \to \mathfrak{Y} \\ Z &: \mathcal{W} \to Z \end{aligned}$$

be random variables on the probability space $(\mathcal{W},\mathcal{B}_{\mathcal{W}},P_{\mathcal{W}})$, let $P_{\mathfrak{X}}$, $P_{\mathfrak{Y}}$, and $P_{\mathfrak{X}\times\mathfrak{Y}}$ be the probabilities on $\mathfrak{X}$, $\mathfrak{Y}$, and $\mathfrak{X} \times \mathfrak{Y}$ induced by X,Y and (X,Y) respectively, and let $Q(x,B)$, $H(x,y;C)$ be stochastic kernels on $(\mathfrak{X},\mathfrak{Y})$, $(\mathfrak{X} \times \mathfrak{Y},Z)$ which satisfy

$$\text{(A.1.14)} \qquad Q(x,A) = P_{\mathcal{W}}[Y \in A|X = x] \qquad \text{a.s. } P_{\mathfrak{X}}$$

for all $B \in \mathcal{B}_{\mathfrak{Y}}$ and

$$\text{(A.1.15)} \qquad H(x,y;B) = P_{\mathcal{W}}[Z \in B|X = x \ , \ Y = y] \qquad \text{a.s. } P_{\mathfrak{X}\times\mathfrak{Y}}$$

Then $J(x,C)$ defined by (A.1.3) satisfies

$$\text{(A.1.6)} \qquad J(x,C) = P_{\mathcal{W}}[(Y,Z) \in C|X = x] \qquad \text{a.s. } P_{\mathfrak{X}}$$

for all $C \in \mathcal{B}_{\mathfrak{Y}} \times \mathcal{B}_Z$.

Proof: From Lemma A.1.5 , $H^*(x,y;C)$ defined by (A.1.4) satisfies

$$H^*(x,y;C) = E_{\mathcal{W}}[\chi_C(Y,Z)|X = x \ , \ Y = y] \qquad \text{a.s. } P_{\mathfrak{X}\times\mathfrak{Y}} \quad .$$

From Lemma A.1.5, (A.1.5) , and the smoothing property for condition expectations

$$J^*(x,y;C) = E_{\mathcal{W}}[E_{\mathcal{W}}[\chi_C(Y,Z)|X,Y]|X = x]$$

$$= E_{\mathcal{W}}[\chi_C(Y,Z)|X = x] = P_{\mathcal{W}}[(Y,Z) \in C|X = x] \qquad \text{a.s. } P_{\mathfrak{X}}$$

Since J^* is the stochastic kernel determined by (A.1.3) , the result follows.

<u>Lemma</u> A.1.8 . Let

$$(\mathcal{W},\mathcal{B}_{\mathcal{W}},P_{\mathcal{W}}) = (\mathcal{W}' \times \mathcal{W}'', \mathcal{B}_{\mathcal{W}'} \times \mathcal{B}_{\mathcal{W}''} , P_{\mathcal{W}'} \times P_{\mathcal{W}''})$$

be a product probability space, let

$$\text{(A.1.17)} \qquad \begin{aligned} &X : \mathcal{W}' \to \mathfrak{X} \\ &\psi : \mathfrak{X} \times \mathcal{W}'' \to \mathcal{Y} \end{aligned}$$

be measureable functions, and let

$$\text{(A.1.18)} \qquad Y(w) = \psi(X(w'),w'') .$$

Then

$$\text{(A.1.19)} \qquad H(x,B) = P_{\mathcal{W}''}(\{w''|\psi(x,w'') \in B\})$$

is a stochastic kernel on $(\mathfrak{X},\mathcal{Y})$ and satisfies

$$\text{(A.1.20)} \qquad H(x,B) = P_{\mathcal{W}}[Y \in B|X = x] \qquad \text{a.s. } P_{\mathfrak{X}}$$

for all $B \in \mathcal{B}_{\mathcal{Y}}$.

Proof: Since w' and w'' are independent and X depends only on w' ,

$$\text{(A.1.21)} \qquad P_{\mathcal{W}}[W'' \in B|X = x] = P_{\mathcal{W}''}(B) \qquad \text{a.s. } P_{\mathfrak{X}}$$

for $B \in \mathcal{B}_{\mathcal{W}''}$ where

$$W''(w) = w''$$
$$X(w) = X(w') \qquad .$$

Since $\chi_B(\psi(x,w''))$ is a stochastic kernel on $(\mathfrak{X} \times \mathcal{W}'',\mathcal{Y})$ from Lemma A.1.2 ii)

$$H(x,B) = \int \chi_B(\psi(x,w''))P_{\mathcal{W}''}(dw'')$$

is a stochastic kernel on $(\mathcal{X},\mathcal{Y})$. Then from Lemma A.1.5, (A.1.21) , and (A.1.18)

$$H(x,B) = E_{\mathcal{W}}[\chi_B(\psi(X,W''))|X = x]$$

$$= E_{\mathcal{W}}[\chi_B(Y)|X = x] = P_{\mathcal{W}}[Y \in B|X = x] \qquad \text{a.s. } P_{\mathcal{X}} \quad .$$

The following lemma is a rather straight forward generalization of example 2.6 p.613 of Doob [4] .

Lemma A.1.9 . Let $J(x,A)$ be a stochastic kernel on $(\mathcal{X},\mathcal{Y}\times\mathcal{Z})$ where $(\mathcal{Y},\mathcal{B}_{\mathcal{Y}})(Z,\mathcal{B}_Z)$ are finite dimensional Euclidean vector spaces with the usual Borel σ-fields, and let

$$\text{(A.1.22)} \qquad J(x,B,C) = J(x,B\times C)$$

for $B \in \mathcal{B}_{\mathcal{Y}}$, $C \in \mathcal{B}_Z$. Then there exists $G(x,y;C)$ a stochastic kernel on $(\mathcal{X} \times \mathcal{Y},Z)$ that satisfies

$$\text{(A.1.23)} \qquad J(x,B,C) = \int_B G(x,y;C)J(x,dy,Z)$$

for all $B \in \mathcal{B}_{\mathcal{Y}}$, $C \in \mathcal{B}_Z$.

Proof: Let $\mathcal{B}^*$ and $\mathcal{C}^*$ be the sub-classes of $\mathcal{B}_{\mathcal{Y}}$ and $\mathcal{B}_Z$ which contain all products of closed, half-infinite, real intervals with rational upper end points. These classes are countable, and they generate their respective σ -fields $\mathcal{B}_{\mathcal{Y}}$ and $\mathcal{B}_Z$. Let

$$\mathcal{B}^* = \{B_1,B_2,\ldots\}$$

and let B_j^n , $j = 1,2,\ldots,2^n$ be intersections of the form

$$\bigcap_{j=1}^{n} (B_j)^{i_j}$$

where $i_j = 0$ or 1 and $(B_j)^1 = B_j$, $(B_j)^0 = B_j^C$ (complement of B_j) . Then for each n the $B_j^{(n)}$, $j = 1,\ldots,2^n$ are disjoint and

$$\text{(A.1.24)} \qquad \sum_j B_j^n = \mathcal{Y} \quad .$$

for $C \in \mathcal{C}^*$, define

$$g^n(x,y,C) = \sum_j \chi_{A_j^n}(x)\, \chi_{B_j^n}(y)\, \frac{J(x,B_j^n,C)}{J(x,B_j^n,Z)} \tag{A.1.25}$$

where

$$A_j^n = \{x \mid J(x,B_j^n,Z) \neq 0\} .$$

The functions $g^n(x,y,C)$ are clearly $\mathcal{B}_\mathcal{X} \times \mathcal{B}_\mathcal{Y}$ measurable for all n and $C \in \mathcal{C}^*$. For x and C fixed $J(x,B,C)$ is a measure in $B \in \mathcal{B}_\mathcal{Y}$. Further, from (A.1.22) it is absolutely continuous with respect to $J(x,\cdot,Z)$. Thus from Theorem 2.5 p. 612 of Doob [4] there exists functions $\bar{g}(x,y,C)$, $\mathcal{B}_\mathcal{Y}$-measurable in y for all $x \in \mathcal{X}$, $C \in \mathcal{C}^*$ that satisfy

$$\lim_{n\to\infty} g^n(x,y,C) = \bar{g}(x,y,C) \qquad \text{a.s. } J(x,\cdot,Z) \tag{A.1.26}$$

and

$$\int_B \bar{g}(x,y,C)J(x,dy,Z) = J(x,B,C) \tag{A.1.27}$$

for all $B \in \mathcal{B}_Z$. Let

$$S^* = \{(x,y) \mid \text{for all } C \in \mathcal{C}^* \text{ the sequence } g^n(x,y,C) \text{ converges}\}$$

and define

$$g^*(x,y,C) = \begin{cases} \lim_{n\to\infty} g^n(x,y,C) & (x,y) \in S^* \\ P^*(C) & (x,y) \notin S^* \end{cases} \tag{A.1.28}$$

where $P^*(C)$ is an arbitrary probability measure on $\mathcal{B}_Z$. Since the $g^n(x,y,C)$ are jointly measurable, $S^* \in \mathcal{B}_\mathcal{X} \times \mathcal{B}_\mathcal{Y}$ and $g^*(x,y,C)$ is jointly measurable. Since $\mathcal{C}^*$ is countable, from (A.1.26) and (A.1.28) for all $x \in \mathcal{X}$ and $C \in \mathcal{C}^*$

$$g^*(x,y,C) = \bar{g}(x,y,C) \qquad \text{a.s. } J(x,\cdot,Z) . \tag{A.1.29}$$

For $C' \subset C''$, $C' \in \mathcal{C}^*$, $C'' \in \mathcal{C}^*$, from (A.1.25)

$$0 \leq g^n(x,y,C') \leq g^n(x,y,C'') \leq 1$$

for all n, x, and y, and thus

(A.1.30) $\quad 0 \leq g^*(x,y,C') \leq g^*(x,y,C'') \leq 1$

Further

$$g^n(x,y,\phi) = 0$$

implies that

(A.1.31) $\quad g^*(x,y,\phi) = 0 \quad .$

From (A.1.27), (A.1.29), (A.1.30) and the properties of $J(x,B,C)$ it can easily be shown that

(A.1.32) $\quad \bar{g}(x,y,Z) = 1 \qquad$ a.s. $J(x,\cdot,Z)$

and $C_n \downarrow C_0$ implies that

(A.1.33) $\quad \bar{g}(x,y,C_n) \downarrow \bar{g}(x,y,C_0) \qquad$ a.s. $J(x,\cdot,Z) \quad .$

Let $\lambda_1,\ldots,\lambda_m$ be the rational upper end points of the set $C \in \mathcal{C}^*$ and let

$$g^*(x,y;\lambda_1,\ldots,\lambda_m) = g^*(x,y,C) \quad .$$

From (A.1.30) $g^*(x,y;\lambda_1,\ldots,\lambda_m)$ is a non-decreasing function of each λ_i. Thus its limits from the right exist at each rational point $(\lambda_1,\ldots,\lambda_m)$. From (A.1.33) and (A.1.29) for each x and almost all y g^* is continuous from the right at each rational points. Let

$$S = \{(x,y) \mid g^*(x,y_j;\lambda_1,\ldots,\lambda_m) \text{ is continuous from the right for all rational points } (\lambda_1,\ldots,\lambda_n) \text{ and } g^*(x,y;Z) = 1\}$$

and for $C \in \mathcal{C}^*$.

(A.1.34) $$g(x,y;C) = \begin{cases} g^*(x,y,C) & (x,j) \in S \; . \\ P^*(C) & (x,y) \notin S \; . \end{cases}$$

Again $g(x,y;C)$ is clearly jointly measurable. Further it is a non-decreasing function, continuous from the right for all rational points and satisfies

$$g(x,y,\phi) = 0$$

$$g(x,y,Z) = 1$$

for all (x,y) . It follows then as in Theorem 9.4 p. 29 of Doob [4] that for each (x,y) , $g(x,y,C)$, $C \in \mathcal{C}^*$ can be extended to $G(x,y,C)$ a probability measure in $C \in \mathcal{B}_Z$ which satisfies

(A.1.35) $$G(x,y,C) = g(x,y,C) \qquad \text{for } C \in \mathcal{C}^* .$$

From (A.1.29), (A.1.32), (A.1.33), and (A.1.34) for all $C \in \mathcal{C}^*$

$$G(x,y,C) = \bar{g}(x,y,C) \qquad \text{a.s. } J(x,\cdot,Z)$$

and hence from (A.1.27), (A.1.23) is satisfied for $C \in \mathcal{C}^*$.

It remains to show that $G(x,y,C)$ is jointly measurable for all $C \in \mathcal{B}_Z$ (from (A.1.35) this holds for $C \in \mathcal{C}^*$) and that (A.1.23) is satisfied for all $C \in \mathcal{B}_Z$. This is accomplished in each case by showing that the class of sets for which the desired property holds is a monotone class and that it contains the field over $\mathcal{C}^*$.

Lemma A.1.10. Let μ be a σ-finite measure on $(\mathcal{Y},\mathcal{B}_{\mathcal{Y}})$, let $j^*(x,y;C)$ be jointly measurable in (x,y) for all C and a measure in $C \in \mathcal{B}_Z$ for all (x,y) , and let

(A.1.36) $$J(x,B \times C) = \int_B j^*(x,y;C)\, \mu(dy)$$

be a stochastic kernel on $(\mathcal{X},\mathcal{Y} \times Z)$. Define G by

(A.1.37) $$G(x,y;C) = \begin{cases} \dfrac{j^*(x,y;C)}{j^*(x,y;Z)} & \text{if } 0 < j^*(x,y;Z) < \infty \\ G^*(C) & \text{otherwise} \end{cases}$$

where G^* is a probability measure on Z . Then $G(x,y;C)$ is a stochastic kernel on $(\mathcal{X} \times \mathcal{Y},Z)$ and satisfies (A.1.23) .

Proof: From the properties of j^* , $G(x,y,C)$ defined by (A.1.37) is clearly a stochastic kernel. For x and C fixed, from (A.1.36) $J_C(B) = J(x,B \times C)$ is absolutely continuous with respect to μ and

(A.1.38) $$\frac{dJ_C}{d\mu}(y) = j^*_C(y) = j^*(x,y,C) \qquad \text{a.e. } \mu$$

From (A.1.36) for $c = Z$ and x fixed

$$j_Z^*(y) < \infty \qquad \text{a.e. } \mu \tag{A.1.39}$$

and

$$J_Z[\{y \mid j_Z^*(y) = 0\}] = 0 \quad .$$

From the absolute continuity of J_Z, the exceptional set in (A.1.38) also has J_Z measure zero. Thus

$$0 < j_Z^*(y) < \infty \qquad \text{a.s. } J_Z \tag{A.1.40}$$

From ex. 21 p. 141 of Loève [12]

$$\frac{dJ_C}{d\mu} = \frac{dJ_C}{dJ_Z} \cdot \frac{dJ_Z}{d\mu} \qquad \text{a.e. } \mu \tag{A.1.41}$$

Thus from (A.1.41), (A.1.38), (A.1.40), and (A.1.37)

$$\frac{dJ_C}{dJ_Z} = \frac{j_C^*(y)}{j_Z^*(y)} = G(x,y,C) \qquad \text{a.s. } J_Z$$

It follows then that

$$\int_B G(x,y,C)J_Z(dy) = J_C(B)$$

and (A.1.23) holds.

Lemma A.1.11. Let $G(y,C)$ and $\bar{G}(y,C)$ be stochastic kernels on $(\mathcal{Y},Z)$ where $(\mathcal{Y},\mathcal{B}_{\mathcal{Y}},P_{\mathcal{Y}})$ is a probability space and Z is a finite dimensional Euclidean space and suppose that

$$G(y,C) = \bar{G}(y,C) \qquad \text{a.s. } P_{\mathcal{Y}} \tag{A.1.42}$$

for all $C \in \mathcal{B}_Z$. Then

$$P_{\mathcal{Y}}[\{y \mid G(y,C) = \bar{G}(y,C) \text{ for all } C \in \mathcal{B}_Z\}] = 1 \ . \tag{A.1.43}$$

Proof: Proceeding as in Lemma A.1.9, since C^* is countable

$$P_{\mathcal{Y}}[\{y \mid G(y,C) = \bar{G}(y,C) \text{ for } C \in C^*\}] = 1 \tag{A.1.44}$$

Since each is a stochastic kernel, for all y $G(y,C)$ and $\bar{G}(y,C)$ are continuous on C^* and satisfy

$$G(y,\emptyset) = \bar{G}(y,\emptyset) = 0$$

$$G(y,Z) = \bar{G}(y,Z) = 1$$

Thus each can be extended uniquely to a probability on $\mathcal{B}_Z$. For y in the set of (A.1.44) these extensions are idendical since each is determined by its values on $\mathcal{C}^*$, and (A.1.43) follows from (A.1.44).

Lemma A.1.12. Let $G(y,A)$ be a stochastic kernel on $(\mathcal{Y},\mathcal{X})$ then

(A.1.45) $\mathcal{G}: \mathcal{Y} \to \mathcal{X}^*$

defined by

(A.1.46) $\mathcal{G}(y) = G(y,\cdot)$

is a measurable transformation, where $\mathcal{X}^*$ is the space of all probability measures on $(\mathcal{X},\mathcal{B}_{\mathcal{X}})$ and $\mathcal{B}_{\mathcal{X}^*}$ is generated by $G(A)$, $A \in \mathcal{B}_{\mathcal{X}}$.

Proof: Let

$$B_{A,c} = \{G(\cdot)\,|\,G(A) \leq c\}$$

where $A \in \mathcal{B}_{\mathcal{X}}$ and c is a real number. Then $\mathcal{B}_{\mathcal{X}^*}$ is the minimal σ-field over the class

$$\mathcal{B} = \{B_{A,c}\,|\,A \in \mathcal{B}_{\mathcal{X}}, \; c \text{ real}\} .$$

For $\mathcal{G}$ defined (A.1.46)

$$\{y\,|\,\mathcal{G}(y) \in B_{A,c}\} = \{y\,|\,G(y,A) \leq c\} \in \mathcal{B}_{\mathcal{Y}}$$

since $G(y,A)$ is a stochastic kernel. Thus from (D') p. 107 of Loève [12] $\mathcal{G}$ is measurable.

Lemma A.1.13. Let

(A.1.47) $\varphi: \mathcal{X} \to Z$

(A.1.48) $\psi: \mathcal{X} \times \mathcal{Y} \to \mathcal{W}$

be measurable and let $Q(A,B)$ and $H(z,C)$ be stochastic kernels on $(\mathfrak{X},\mathfrak{Y})$ and $(Z,\mathfrak{Z})$ which satisfy

$$H(\varphi(x),C) = Q(x,\{y|\psi(x,y) \in C\}) . \tag{A.1.49}$$

Then for $f(x,w)$ measurable and non-negative

$$\int f(x,w)H(\varphi(x),dw) = \int f(x,\psi(x,y))Q(x,dy) \tag{A.1.50}$$

Proof: For x fixed from (A.1.49) $H(\varphi(x,C)$ is the distribution induced by the random variable $w = \psi(x,y)$ on the probability space $(\mathfrak{Y},\mathfrak{B}_{\mathfrak{Y}},Q(x,B))$. The result then follows from Lemma 2 p. 38 of Lehman [11] .

A.2 P-<u>ess inf</u>

The P-ess inf for a family of integrable functions is defined by Dunford and Schwartz [6] . That definition is extended in Theorem A.2.1 to families of a.s. non-negative,measurable functions. Elementary properties of the P-ess inf are developed in Lemmas A.2.6-A.2.14. In Theorems A.2.2 and A.2.3 properties relating to conditional expectations are proved. These are crucial as denomstrating the submartingale property of the conditional loss function $\hat{F}_t$ in section 4.2.

<u>Definition</u> A.2.1. Let $\{f_\gamma(\omega)\}$, $\gamma \in C$ be a family of measurable, a.s. non-negative (possibly $+\infty$) functions on a probability space $(\Omega,\mathfrak{A},P)$. Then

$$f(\omega) = \text{P-ess inf}_{\gamma \in C} \ f_\gamma(\omega) \qquad \text{a.s. } P \tag{A.2.1}$$

provided

i) $f(\omega)$ is $\mathfrak{A}$-measurable;

ii) for all $\gamma \in C$

$$f(\omega) \leq f_\gamma(\omega) \qquad \text{a.s. } P \ ;$$

iii) if $g(\omega)$ satisfies i) and ii) , then

$$g(\omega) \leq f(\omega) \qquad \text{a.s. } P \ .$$

<u>Lemma</u> A.2.1. If f and f' satisfy i)-iii) of Definition A.2.1, then

$$0 \leq f(\omega) = f(\omega') \qquad \text{a.s. } P \ . \tag{A.2.2}$$

Proof: Since the functions $f_\gamma(\omega)$ are a.s. non-negative

$$0 \leq f_\gamma(\omega) \qquad \text{a.s. } P$$

for all $\gamma \in C$. Thus from iii) of Definition A.2.1 for $g(\omega) = 0$

$$0 \leq f(\omega) \qquad \text{a.s. } P \ .$$

Since $f'(\omega)$ satisfies i) and ii), from iii) for $f(\omega)$ with $g(\omega) = f'(\omega)$,

$$f'(\omega) \leq f(\omega) \qquad \text{a.s. } P \ .$$

Similarly

$$f(\omega) \leq f'(\omega) \qquad \text{a.s. } P \ ,$$

and the result (A.2.2) follows.

Lemma A.2.2. Let $\{\nu_\gamma\}$, $\gamma \in C$, be a family of measures on $(\Omega,\mathfrak{A})$ which contains at least one finite member. For $A \in \mathfrak{A}$, define

$$\nu(A) = \inf_{\substack{\gamma_1,\ldots,\gamma_n \\ A_1,\ldots,A_n \ni \\ \gamma_i \in C, A_i \in \mathfrak{A}, \ \sum_i A_i = A}} \sum_{i=1}^{n} \nu_{\gamma_i}(A_i) \ . \tag{A.2.3}$$

Then $\nu(A)$ is a finite, non-negative, finitely additive set function on $(\Omega,\mathfrak{A})$.

Proof: Since the $\nu_{\gamma_i}(A_i)$ are non-negative, $\nu(A)$ is clearly non-negative. By assumption there exist $\nu_{\gamma_0}(\cdot)$ finite. Thus

$$\nu(\Omega) \leq \nu_{\gamma_0}(\Omega) < \infty \ ,$$

and $\nu(\cdot)$ is finite. Take $A_\alpha \in \mathfrak{A}$ such that

$$\sum_{\alpha=1}^{m} A_\alpha = A \ . \tag{A.2.4}$$

For $\epsilon > 0$ and each α, take $\gamma_{\alpha,1},\gamma_{\alpha,2},\ldots,\gamma_{\alpha,n_\alpha}$ and $A_{\alpha,1},\ldots,A_{\alpha,n_\alpha}$ such that

$$A_\alpha = \sum_{i=1}^{n_\alpha} A_{\alpha,i} \tag{A.2.5}$$

and

$$(A.2.6) \qquad \sum_i \nu_{\alpha i}(A_{\alpha i}) \leq \nu(A_\alpha) + \frac{\varepsilon}{m} \quad .$$

From (A.2.4) and (A.2.5)

$$\sum_{\alpha=1}^{m} \sum_{i=1}^{n_\alpha} A_{\alpha,i} = A \quad .$$

Thus from (A.2.3) and (A.2.6)

$$(A.2.7) \qquad \nu(A) \leq \sum_{\alpha=1}^{m} \sum_{i=1}^{n_\alpha} \nu_{\alpha,i}(A_{\alpha,i}) \leq \sum_{\alpha=1}^{m} \nu(A_\alpha) + \varepsilon \quad .$$

Take $\gamma_1,\ldots,\gamma_n$, $A_1',\ldots,A_n'$ such that

$$(A.2.8) \qquad \sum_{i=1}^{m} A_i' = A$$

and

$$(A.2.9) \qquad \sum_{i=1}^{n} \nu_{\gamma_i'}(A_i') \leq \nu(A) + \varepsilon \ .$$

From (A.2.8) and (A.2.4) for all α

$$(A.2.10) \qquad \sum_{i=1}^{n} (A_i' \cap A_\alpha) = A \cap A_\alpha = A_\alpha$$

and for all i

$$(A.2.11) \qquad \sum_{\alpha=1}^{m} (A_i' \cap A_\alpha) = A_i' \cap A = A_i' \quad .$$

From (A.2.10) and (A.2.3)

$$(A.2.12) \qquad \nu(A_\alpha) \leq \sum_{i=1}^{n} \nu_{\gamma_i}(A_i' \cap A_\alpha) \quad .$$

Thus, from (A.2.12), (A.2.11), and (A.2.9)

$$(A.2.13) \qquad \sum_{\alpha=1}^{m} \nu(A_\alpha) \leq \sum_{i=1}^{n} \sum_{\alpha=1}^{m} \nu_{\gamma_i}(A_i' \cap A_\alpha) = \sum_{i=1}^{n} \nu_{\gamma_i}(A_i') \leq \nu(A) + \varepsilon \quad .$$

The result follows from (A.2.7) and (A.2.13).

Lemma A.2.3. Let $\{\nu_\gamma\}$, $\gamma \in C$, be a family of measures on $(\Omega,\mathfrak{A})$ that contains at least one finite member. Then $\nu(\cdot)$ defined by (A.2.3) is a finite measure on $(\Omega,\mathfrak{A})$.

Proof: Take $A_n \downarrow \emptyset$ and $\gamma \in C$, then from (A.2.3) and the continuity of the ν_γ

$$0 \leq \nu(A_n) \leq \nu_\gamma(A_n) \downarrow 0 \quad .$$

Thus ν is finite, finitely additive and continuous at $\emptyset$. From A. p. 84 of Loève [12], it follows that ν is a measure.

<u>Lemma</u> A.2.4. Let $\{f_\gamma(\omega)\}$, $\gamma \in C$ be a family of measurable, a.s. non-negative (possibly $+\infty$) funtions on a probability space $(\Omega,\mathfrak{A},P)$, let the family contain at least one integrable member, and let

(A.2.14) $$\nu_\gamma(A) = \int_A f_\gamma(\omega)dP$$

for $A \in \mathfrak{A}$. Then $\{\nu_\gamma\}$, $\gamma \in C$ is a family of measures on $\{\Omega,\mathfrak{A}\}$ which contains at least one finite member, ν defined by (A.2.3) is a finite measure absolutely continuous with respect to P, and

(A.2.15) $$\frac{d\nu}{dP} = \text{P-ess}\inf_{\gamma \in C} f_\gamma(\omega) \qquad \text{a.s. } P \quad .$$

Proof: Since $\{f_\gamma(\omega)\}$, $\gamma \in C$, contains an integrable member γ_0,

$$\nu_{\gamma_0}(\Omega) = \int_\Omega f_{\gamma_0}(\omega)dP < \infty \quad ,$$

and ν_{γ_0} is finite. From Lemma A.2.3 ν is a finite measure. Take $A \in \mathfrak{A}$ such that

$$P(A) = 0 \quad .$$

Then from (A.2.3) and (A.2.14) for $\gamma \in C$

$$0 \leq \nu(A) \leq \nu_\gamma(A) = 0 \quad .$$

This ν is absolutely continuous with respect to P. Since the Radon-Nikodym derivative is $\mathfrak{A}$-measurable,

(A.2.16) $$f = \frac{d\nu}{dP}$$

satisfies i) of Definition A.2.1. From (A.2.3) for all $\gamma \in C$ and $A \in \mathfrak{A}$

$$\nu(A) \leq \nu_\gamma(A) \; .$$

It follows then from (A.2.14) that

$$\frac{d\nu}{dP} \leq \frac{d\nu_\gamma}{dP} = f_\gamma(\omega) \qquad \text{a.s. } P \text{ .}$$

Thus f given by (A.2.16) satisfies ii) of Definition A.2.1. Let g satisfy i) and ii) , then

$$\int_A g\,dP \leq \nu_\gamma(A) \tag{A.2.17}$$

for all $\gamma \in C$ and $A \in \mathfrak{A}$. For $A \in \mathfrak{A}$ and $\epsilon > 0$, take $\gamma_1,\ldots,\gamma_n$, $A_1,\ldots,A_n$ such that

$$\sum_{i=1}^{n} A_i = A \tag{A.2.18}$$

and

$$\sum_{i=1}^{n} \nu_{\gamma_i}(A_i) \leq \nu(A) + \epsilon \text{ .} \tag{A.2.19}$$

Then from (A.2.18), (A.2.17), and (A.2.19)

$$\int_A g\,dP = \sum_{i=1}^{n} \int_{A_i} g\,dP \leq \sum_{i=1}^{n} \nu_{\gamma_i}(A_i) \leq \nu(A) + \epsilon \text{ .}$$

Thus

$$\int_A g\,dP \leq \nu(A)$$

for all $A \in \mathfrak{A}$, it follows that

$$g \leq \frac{d\nu}{dP} \qquad \text{a.s. } P \text{ ,}$$

and hence iii) of Definition A.2.1 holds.

<u>Lemma</u> A.2.5. Let $\{f_\gamma(\omega)\}$, $\gamma \in C$ be a family of measurable, a.s. non-negative (possibly $+\infty$) functions on a probability space $(\Omega,\mathfrak{A},P)$, for N a positive integer let

$$f_\gamma^N(\omega) = \begin{cases} f_\gamma(\omega) & \text{if } f_\gamma(\omega) \leq N \\ N & \text{if } f_\gamma(\omega) > N \text{ ,} \end{cases} \tag{A.2.20}$$

and let

$$f^N(\omega) = P\text{-ess}\inf_{\gamma \in C} f_\gamma^N \qquad \text{a.s. } P \text{ .} \tag{A.2.21}$$

Then there exists $f(\omega)$ such that

(A.2.22) $$f^N_{(\omega)} \uparrow f(\omega) \qquad \text{a.s. } P \quad ,$$

and

(A.2.23) $$f(\omega) = P\text{-ess}\inf_{\gamma \in C} f_\gamma(\omega) \quad .$$

Proof: The functions $\{f^N_\gamma\}$, $\gamma \in C$ are a.s. non-negative and integrable. Thus from Lemma A.2.4 there exists $f^N(\omega)$ which satisfies (A.2.21). From ii) of Definition A.2.1 and (A.2.20) , for all $\gamma \in C$

$$f^N(\omega) \leq f^N_\gamma(\omega) \leq f^{N+1}_\gamma(\omega) \qquad \text{a.s. } P \quad .$$

Thus from iii) of Definition A.2.1 for the family $\{f^{N+1}_\gamma\}$, $\gamma \in C$

$$f^N(\omega) \leq f^{N+1}(\omega) \qquad \text{a.s. } P \quad .$$

Thus $f^N(\omega)$ is increasing a.s., and there exists $f(\omega)$ measurable which satisfies (A.2.22). For each $\gamma \in C$, from (A.2.20)

(A.2.24) $$f^N_\gamma \uparrow f_\gamma \quad .$$

Thus from ii) of Definition A.2.1 for $\{f^N_\gamma\}$ and (A.2.24)

$$f^N \leq f^N_\gamma \leq f_\gamma \qquad \text{a.s. } P$$

and from (A.2.22)

$$f(\omega) \leq f_\gamma(\omega) \qquad \text{a.s. } P \quad .$$

Thus f satisfies ii) of Definition A.2.1. Let g be measurable and satisfy

$$g \leq f_\gamma \qquad \text{a.s. } P$$

for all $\gamma \in C$. Then g^N defined by the truncation procedure (A.2.20) is measurable and satisfies

$$g^N \leq f^N_\gamma \qquad \text{a.s. } P \quad .$$

Thus from iii) of Definition A.2.1 for $\{f_\gamma^N\}$, $\gamma \in C$, (A.2.21) and (A.2.22)

$$g^N \leq f^N \leq f ,$$

and hence

$$g = \lim_{N\to\infty} g^N \leq f .$$

Thus f defined by (A.2.22) also satisfies iii) of Definition A.2.1.

Theorem A.2.1. For $\{f_\gamma(\omega)\}$, $\gamma \in C$, a family of measurable, a.s. non-negative (possibly $+\infty$) functions on a probability space $(\Omega,\mathfrak{A},P)$, the P-ess inf f_γ exists and is a.s. non-negative and a.s. unique.

Proof: Existence follows from Lemma A.2.5, and a.s. non-negativity and uniqueness from Lemma A.2.1.

Lemma A.2.6. If

$$\text{(A.2.25)} \qquad f_\gamma(\omega) = k_\gamma \qquad \text{a.s. } P$$

for all $\gamma \in C$ where k_γ are non-negative (possibly $+\infty$) constants, then

$$\text{(A.2.26)} \qquad \text{P-ess}\inf_{\gamma\in C} f_\gamma(\omega) = \inf_{\gamma\in C} k_\gamma \qquad \text{a.s. } P .$$

Proof: Let

$$k_0 = \inf_{\gamma\in C} k_\gamma .$$

Then from (A.2.25) , for all γ

$$k_0 \leq k_\gamma = f_\gamma(\omega) \qquad \text{a.s. } P .$$

Take γ_1 , $\gamma_2, \ldots$ such that

$$k_{\gamma_n} \downarrow k_0 .$$

Then $g(\omega)$ measurable and satisfying

$$g(\omega) \leq f_{\gamma_n}(\omega) \qquad \text{a.s. } P$$

implies from (A.2.25) that

$$g(\omega) \leq k_{\gamma_n} \qquad n = 1,2,\ldots \qquad \text{a.s. } P$$

and hence that

$$g(\omega) \leq k_0 \qquad \text{a.s. } P \quad .$$

<u>Lemma</u> A.2.7. If the family of measurable, a.s. non-negative (possibly $+\infty$) functions $\{f_\gamma(\omega)\}$ on $(\Omega,\mathfrak{A},P)$ has the single member f_{γ_0} or all members are a.s. equal to f_{γ_0}, then

(A.2.27) $$\underset{\gamma \in C}{P\text{-ess inf}}\ f_\gamma(\omega) = f_{\gamma_0}(\omega) \qquad \text{a.s. } P \quad .$$

Proof: The proof of this lemma is trivial.

<u>Lemma</u> A.2.8. Let $\{f_\gamma(\omega)\}$ and $\{g_\gamma(\omega)\}$, $\gamma \in C$, be families of measurable, a.s. non-negative (possibly $+\infty$) functions on $(\Omega,\mathfrak{A},P)$ which satisfy

(A.2.28) $$g_\gamma(\omega) \leq f_\gamma(\omega) \qquad \text{a.s. } P$$

for all $\gamma \in C$. Then

(A.2.29) $$\underset{\gamma \in C}{P\text{-ess inf}}\ g_\gamma(\omega) \leq \underset{\gamma \in C}{P\text{-ess inf}}\ f_\gamma(\omega) \qquad \text{a.s. } P \quad .$$

Proof: Let

$$g(\omega) = \underset{\gamma \in C}{P\text{-ess inf}}\ g_\gamma(\omega) \qquad \text{a.s. } P \quad .$$

Then from ii) of Definition A.2.1 for $\{g_\gamma\}$ and (A.2.28), for all $\gamma \in C$

$$g(\omega) \leq g_\gamma(\omega) \leq f_\gamma(\omega) \qquad \text{a.s. } P \ .$$

Thus from iii) of Definition A.2.1 for $\{f_\gamma\}$

$$g(\omega) \leq \underset{\gamma \in C}{P\text{-ess inf}}\ f_\gamma(\omega) \qquad \text{a.s. } P \quad .$$

<u>Lemma</u> A.2.9. Let $\{f_\gamma(\omega)\}$ and $\{g_\gamma(\omega)\}$, $\gamma \in C$ be families of measurable, a.s. non-negative (possibly $+\infty$) functions on $(\Omega,\mathfrak{A},P)$ which satisfy

(A.2.30) $$g_\gamma(\omega) = f_\gamma(\omega) \qquad \text{a.s. } P \quad .$$

Then

$$\text{(A.2.31)} \qquad \underset{\gamma}{\text{P-ess inf}}\ g_\gamma(\omega) = \underset{\gamma}{\text{P-ess inf}}\ f_\gamma(\omega) \quad \text{a.s. P} .$$

Proof: This follows from Lemma A.2.8.

Lemma A.2.10. Let $\{f_\gamma(\omega)\}$, $\gamma \in C$ be a family of measurable, a.s. non-negative (possibly $+\infty$) functions on $(\Omega,\mathfrak{A},P)$, and let $C^* \subset C$. Then

$$\text{(A.2.32)} \qquad \underset{\gamma\in C}{\text{P-ess inf}}\ f_\gamma(\omega) \leq \underset{\gamma\in C^*}{\text{P-ess inf}}\ f_\gamma(\omega) \qquad \text{a.s. P} .$$

Proof: Let

$$f(x) = \underset{\gamma\in C}{\text{P-ess inf}}\ f_\gamma(\omega) \qquad \text{a.s. P} .$$

Then from ii) of Definition A.2.1 for $\{f_\gamma\}$, $\gamma \in C$

$$f(x) \leq f_\gamma(\omega) \qquad \text{a.s. P}$$

for $\gamma \in C$ and hence for $\gamma \in C^*$. Thus from iii) of Definition A.2.1 for $\{f_\gamma\}$, $\gamma \in C^*$

$$f(x) \leq \underset{\gamma\in C^*}{\text{P-ess inf}}\ f_\gamma(\omega) \qquad \text{a.s. P} .$$

Lemma A.2.11. Let $\{f_\gamma(\omega)\}$, $\gamma \in C$, be a family of measurable, a.s. non-negative (possibly $+\infty$) functions on $(\Omega,\mathfrak{A},P)$, and let $b(\omega)$ be an a.s. non-negative (possibly $+\infty$), measurable function. Then

$$\text{(A.2.33)} \qquad \underset{\gamma\in C}{\text{P-ess inf}}\ [f_\gamma(\omega) + b(\omega)] = b(\omega) + \underset{\gamma\in C}{\text{P-ess inf}}\ f_\gamma(\omega) \quad \text{a.s. P} .$$

Proof: From the assumptions of the lemma it is clear that $\{f_\gamma + b\}$, $\gamma \in C$ is a family of measurable, a.s. non-negative (possibly $+\infty$) functions on $(\Omega,\mathfrak{A},P)$. Thus the left side of (A.2.33) is well-defined. Let

$$f(\omega) = \underset{\gamma\in C}{\text{P-ess inf}}\ f_\gamma(\omega) \qquad .$$

It will be shown that $f(\omega) + b(\omega)$ satisfies conditions i)-iii) of Definition A.2.1 for the family $\{f_\gamma + b\}$, $\gamma \in C$. Clearly $f(\omega) + b(\omega)$ is measurable. From ii) of Definition A.2.1 for $\{f_\gamma\}$

$$f(\omega) \leq f_\gamma(\omega) \qquad \text{a.s. P}$$

for all $\gamma \in C$. Thus

$$f(\omega) + b(\omega) \leq f_\gamma(\omega) + b(\omega) \qquad \text{a.s. } P$$

for all $\gamma \in C$. Let $g(\omega)$ be measurable and satisfy

$$g(\omega) \leq f_\gamma(\omega) + b(\omega) \qquad \text{a.s. } P$$

for all $\gamma \in C$ and let

$$B = \{\omega | b(\omega) < \infty\} \quad .$$

Then clearly

$$\chi_B(\omega)[g(\omega) - b(\omega)] < f_\gamma(\omega) \qquad \text{a.s. } P$$

for all $\gamma \in C$. It follows then from the definition of $f(\omega)$ that

$$\chi_B[g(\omega) - b(\omega)] \leq f(\omega) \qquad \text{a.s. } P \text{ ,}$$

and from the definition of B that

$$g(\omega) \leq f(\omega) + b(\omega) \qquad \text{a.s. } P \quad .$$

<u>Lemma</u> A.2.12. Let $\{f_\gamma(\omega)\}$, $\gamma \in C$, be a family of measurable, a.s. non-negative function in $(\Omega,\mathfrak{A},P)$, and let $a(\omega)$ be non-negative, measurable and finite. Then

$$\text{(A.2.34)} \qquad \underset{\gamma \in C}{\text{P-ess inf}} \, [a(\omega)f_\gamma(\omega)] = a(\omega) \, \underset{\gamma \in C}{\text{P-ess inf}} \, f_\gamma(\omega) \qquad \text{a.s. } P$$

where the convention $0 \cdot \infty = 0$ is observed.

Proof: Clearly $\{a(\omega)f_\gamma(\omega)\}$, $\gamma \in C$, is a family of measurable, a.s. non-negative functions. Let

$$f(\omega) = \underset{\gamma}{\text{P-ess inf}} \, f_\gamma(\omega) \qquad \text{a.s. } P$$

$$f'(\omega) = \underset{\gamma}{\text{P-ess inf}} \, a(\omega)f_\gamma(\omega) \qquad \text{a.s. } P \quad .$$

From ii) of Definition A.2.1 for $\{f_\gamma(\omega)\}$

$$f(\omega) \leq f_\gamma(\omega) \qquad \text{a.s. } P$$

for all γ, and hence

$$a(\omega)f(\omega) \leq a(\omega)f_\gamma(\omega) \qquad \text{a.s. } P \ .$$

Thus from iii) of Definition A.2.1 for the family $\{a(\omega)f_\gamma(\omega)\}$

$$a(\omega)f(\omega) \leq f'(\omega) \qquad \text{a.s. } P \ . \tag{A.2.35}$$

Let $\epsilon > 0$, then from ii) for $\{af_\gamma\}$

$$f'(\omega) \leq a(\omega)f_\gamma(\omega) \leq [a(\omega)+\epsilon]f_\gamma(\omega) \qquad \text{a.s. } P$$

for all γ since $f_\gamma(\omega)$ is a.s. non-negative. Thus, since $a(\omega)+\epsilon$ is non-negative and finite

$$\frac{f'(\omega)}{[a(\omega)+\epsilon]} \leq f_\gamma(\omega) \qquad \text{a.s. } P$$

and from iii) for $\{f_\gamma\}$

$$\frac{f'(\omega)}{[a(\omega)+\epsilon]} \leq f(\omega) \qquad \text{a.s. } P \ .$$

Thus

$$f'(\omega) \leq [a(\omega)+\epsilon]f(\omega) \qquad \text{a.s. } P \tag{A.2.36}$$

for all $\epsilon > 0$. Let

$$B = \{\omega \mid a(\omega) \neq 0\} \ . \tag{A.2.37}$$

Then letting $\epsilon_n \to 0$

$$\chi_B(\omega)[a(\omega)+\epsilon_n]f(\omega) \to \chi_B(\omega)a(\omega)f(\omega) \ . \tag{A.2.38}$$

Thus from (A.2.36) and (A.2.38)

$$\chi_B(\omega)f'(\omega) \leq \chi_B(\omega)a(\omega)f(\omega) \qquad \text{a.s. } P \ . \tag{A.2.39}$$

Let

$$f_0(\omega) = \begin{cases} \infty & \omega \in B \\ 0 & \omega \notin B \end{cases} \ . \tag{A.2.40}$$

Then from the convention $0 \cdot \infty = 0$

$$a(\omega)f_\gamma(\omega) \leq f_0(\omega) \qquad \text{a.s. } P$$

for all γ . It follows from Lemmas A.2.8 and A.2.7 that

$$f'(\omega) \leq \text{P-ess inf} \quad f_0\omega = f_0(\omega) \qquad \text{a.s. } P \;.$$

Thus from (A.2.40), (A.2.37), and the convertion $0 \cdot \infty = 0$

$$(A.2.41) \qquad f'(\omega)\chi_{B^c}(\omega) = 0 = \chi_{B^c}(\omega)a(\omega)f(\omega) \qquad \text{a.s. } P$$

it follows from (A.2.39) and (A.2.41) that

$$f'(\omega) < a(\omega)f(\omega) \qquad \text{a.s. } P \;,$$

and hence from (A.3.35) that

$$f'(\omega) = a(\omega)f(\omega) \qquad \text{a.s. } P \;.$$

Lemma A.2.13. Let $\{f_\gamma(y)\}$, $\gamma \in C$ be a family of measurable, non-negative (possibly $+\infty$) funtions on the measurable space $(\mathfrak{Y},\mathfrak{B})$, let

$$(A.2.42) \qquad Y: (\Omega.\mathfrak{A}) \to (\mathfrak{Y},\mathfrak{B})$$

be a measurable transformation on the probability space $(\Omega,\mathfrak{A},P)$, define the measure $Y^{-1}P$ on $(\mathfrak{Y},\mathfrak{B})$ by

$$(A.2.43) \qquad (Y^{-1}P)(B) = P(Y^{-1}B) \;,$$

and let

$$(A.2.44) \qquad f(y) = Y^{-1}P \text{ -ess inf}_{\gamma \in C} \quad f_\gamma(y) \qquad \text{a.s. } Y^{-1}P \;.$$

Then

$$(A.2.45) \qquad \text{P-ess inf}_{\gamma \in C} \quad f_\gamma(Y(\omega)) = f(Y(\omega)) \qquad \text{a.s. } P \;.$$

Proof: Let

$$(A.2.46) \qquad f'(\omega) = \text{P-ess inf}_{\gamma \in C} \quad f_\gamma(Y(\omega)) \qquad \text{a.s. } P \;.$$

From ii) for $\{f_\gamma(y)\}$ and (A.2.44), for all $\gamma \in C$

$$f(y) \leq f_\gamma(y) \qquad \text{a.s. } Y^{-1}P \;.$$

Thus from (A.2.43)

$$f(Y(\omega)) \leq f_\gamma(Y(\omega)) \qquad \text{a.s. } P \;,$$

and from iii) for $\{f_\gamma(Y(\omega))\}$ and (A.2.46)

$$f(Y(\omega)) \leq f'(\omega) \qquad \text{a.s. } P \; .$$

Let $\varphi(\omega)$ be non-negative, measurable and bounded. Then

$$E[\varphi(\omega) | Y = y] = a(y)$$

may be taken non-negative, measurable , and finite. From (A.2.46), for all $\gamma \in C$

$$\varphi(\omega)f'(\omega) \leq \varphi(\omega)f_\gamma(Y(\omega)) \qquad \text{a.s. } P \; .$$

Taking conditional expectations,

$$E[\varphi(\omega)f'(\omega) | Y = y] \leq E[\varphi(\omega)f_\gamma(Y(\omega)) | Y = y] = a(y)f_\gamma(y) \qquad \text{a.s. } Y^{-1}P \; .$$

Thus from iii) for the family $\{a(y)f_\gamma(y)\}$ and Lemma A.2.12,

$$E[\varphi(\omega)f'(\omega) | Y = y] \leq Y^{-1}P\text{-ess}\inf_{\gamma \in C} \; a(y)f_\gamma(y)$$

$$= a(y)Y^{-1}\; P\text{-ess}\inf_{\gamma \in C} \; f_\gamma(y) = f(y)E[\varphi(\omega) | Y = y]$$

$$= E[f(Y(\omega))\varphi(\omega) | Y=y] \qquad \text{a.s. } Y^{-1}P \; .$$

Since this holds for all $\varphi(\omega) = \chi_A(\omega)$, $A \in \mathfrak{A}$, it follows easily that

$$f'(\omega) \leq f(Y(\omega)) \qquad \text{a.s. } P \; .$$

Thus

$$f'(\omega) = f(Y(\omega)) \qquad \text{a.s. } P \; .$$

<u>Definition</u> A.2.2. A family $\{f_\gamma\}$, $\gamma \in C$, of measurable functions on $(\Omega,\mathfrak{A},P)$ has the <u>finite</u> <u>ε-lattice property</u> provided for $\varepsilon > 0$, $\gamma_1 \in C$, and $\gamma_2 \in C$ there exists $\gamma_0 \in C$ that satisfies

$$\text{(A.2.47)} \qquad f_{\gamma_0}(\omega) \leq f_{\gamma_i}(\omega) + \varepsilon \qquad \text{a.s. } P$$

for $i = 1,2$.

<u>Definition</u> A.2.3. A family $\{f_\gamma\}$, $\gamma \in C$, of measurable functions on $(\Omega,\mathfrak{A},P)$ has the <u>countable</u> <u>ε-lattice property</u> provided for $\varepsilon > 0$ and $\gamma_i \in C$, $i = 1,2,\ldots$, there exists $\gamma_0 \in C$ which satisfies

(A.2.48) $$f_{\gamma_0}(\omega) \leq f_{\gamma_i}(\omega) + \varepsilon \qquad \text{a.s. } P$$

for $i = 1,2,\ldots$.

<u>Lemma</u> A.2.14. Let $\{f_\gamma\}$, $\gamma \in C$, be a family of measurable, a.s. non-negative (possibly $+\infty$) functions on $(\Omega,\mathfrak{A},P)$, let the family contain at least one integrable member, and let the family have the finite ε-lattice property. Then $\nu(A)$ defined by (A.2.14) and (A.2.3) satisfies

(A.2.49) $$\nu(A) = \inf_{\gamma \in C} \nu_\gamma(A) \quad .$$

Proof: From (A.2.3)

(A.2.50) $$\nu(A) \leq \inf_{\gamma \in C} \nu_\gamma(A) \quad .$$

For $A \in \mathfrak{A}$ and $\varepsilon > 0$, take $\gamma_1,\ldots,\gamma_n$, $A_1,\ldots,A_n$ such that $\gamma_i \in C$, $A_i \in \mathfrak{A}$,

(A.2.51) $$\sum_{i=1}^{n} A_i = A ,$$

and

(A.2.52) $$\sum_{i=1}^{n} \nu_{\gamma_i}(A_i) \leq \nu(A) + \frac{\varepsilon}{2} .$$

From the finite ε-lattice property there exists $\gamma_0 \in C$ that satisfies

$$f_{\gamma_0} \leq f_{\gamma_i} + \frac{\varepsilon}{2} \qquad \text{a.s. } P$$

for $i = 1,2,\ldots,n$. Thus from (A.2.14)

(A.2.53) $$\nu_{\gamma_0}(A_i) \leq \nu_{\gamma_i}(A_i) + \frac{\varepsilon}{2} P(A_i)$$

for $i = 1,2,\ldots,n$. From (A.2.51), (A.2.53), and (A.2.52)

(A.2.54)
$$\begin{aligned}\inf_{\gamma \in C} \nu_\gamma(A) &\leq \nu_{\gamma_0}(A) = \sum_{i=1}^{n} \nu_{\gamma_0}(A_i) \leq \sum_{i=1}^{n} \nu_{\gamma_i}(A_i) + \frac{\varepsilon}{2} \sum_{i=1}^{n} P(A_i) \\ &\leq \nu(A) + \frac{\varepsilon}{2} + \frac{\varepsilon}{2} P(A) \leq \nu(A) + \varepsilon\end{aligned}$$

Thus (A.2.49) follows from (A.2.50) and (A.2.54) .

<u>Theorem</u> A.2.2. Let $\{f_\gamma(\omega)\}$, $\gamma \in C$, be a family of measurable, a.s. non-negative (possibly $+\infty$) functions on $(\Omega,\mathfrak{A},P)$, let the family contain at least one integrable member, let the family have the finite ϵ-lattice property, let $\mathfrak{B} \subset \mathfrak{A}$ and let $P_{\mathfrak{B}}$ be the restriction of P to the σ-field $\mathfrak{B}$. Then

$$(A.2.55) \qquad E^{\mathfrak{B}}[P\text{-ess}\inf_{\gamma\in C} f_\gamma] = P_{\mathfrak{B}}\text{-ess}\inf_{\gamma\in C} E^{\mathfrak{B}}[f_\gamma] \qquad \text{a.s. } P_{\mathfrak{B}} \ .$$

Proof: It can easily be shown that $\{E^{\mathfrak{B}}[f_\gamma]\}$, $\gamma \in C$, is a family of measurable, a.s. non-negative (possibly $+\infty$) functions on $(\Omega,\mathfrak{B},P)$, that the family contains at least one integrable member and that the family has the finite ϵ-lattice property. Let

$$\nu_\gamma(A) = \int_A f_\gamma \, dP \qquad A \in \mathfrak{A}$$

$$\nu'_\gamma(B) = \int_B E^{\mathfrak{B}}[f_\gamma]dP \qquad B \in \mathfrak{B}$$

$$\nu(A) = \inf_{\gamma\in C} \nu_\gamma(A) \qquad A \in \mathfrak{A}$$

$$\nu'(B) = \inf_{\gamma\in C} \nu'_\gamma(B) \qquad B \in \mathfrak{B} \ .$$

From the properties of conditional expectations, for $B \in \mathfrak{B}$

$$\nu_\gamma(B) = \nu'_\gamma(B)$$

and hence

$$(A.2.56) \qquad \nu(B) = \nu'(B) \quad .$$

Thus from (A.2.15) of Lemma A.2.4 applied to the family $\{E^{\mathfrak{B}}[f_\gamma]\}$, (A.2.49) of Lemma A.2.14 , (A.2.56), and Lemma A.2.4 for $\{f_\gamma\}$, for all $B \in \mathfrak{B}$

$$\int_B \{P_{\mathfrak{B}}\text{-ess}\inf_\gamma E^{\mathfrak{B}}f_\gamma\}dP = \nu'(B) = \nu(B)$$

$$= \int_B [P\text{-ess}\inf_\gamma f_\gamma]dP \quad .$$

Since $P_{\mathfrak{B}}$ -ess inf $E^{\mathfrak{B}}f_\gamma$ is $\mathfrak{B}$-measurable, (A.2.55) follows from the definition of the conditional expectation.

Lemma A.2.15. Let the family $\{f_\gamma(\omega)\}$, $\gamma \in C$, of measurable, a.s. non-negative (possibly $+\infty$) functions on $(\Omega,\mathfrak{A},P)$ have the countable ε-lattice property. Then there exists $\gamma_n \in C$ such that

$$\inf_n f_{\gamma_n}(\omega) = \text{P-ess}\inf_{\gamma \in C} f_\gamma(\omega) \qquad \text{a.s. } P . \tag{A.2.57}$$

Proof: First, the property will be shown for $\{f_\gamma(\omega)\}$ integrable. Take $\gamma_n \in C$ such that

$$\lim_{n\to\infty} \int_\Omega f_{\gamma_n}(\omega)P(d\omega) = \inf_{\gamma \in C} \int_\Omega f_\gamma(\omega)P(d\omega) \quad . \tag{A.2.58}$$

Let

$$g(\omega) = \inf_n f_{\gamma_n}(\omega) \quad . \tag{A.2.59}$$

Then from (A.2.58) and (A.2.59)

$$\int_\Omega g(\omega)P(d\omega) \leq \inf_{\gamma \in C} \int_\Omega f_\gamma(\omega)P(d\omega) \quad . \tag{A.2.60}$$

For $\gamma \in C$ and $\varepsilon > 0$ fixed, let

$$A = \{\omega | f_\gamma(\omega) \leq g(\omega) - \varepsilon\} \; . \tag{A.2.61}$$

It will be shown by contradiction that $P(A) = 0$. Assume $P(A) > 0$, then from the ε-lattice property there exists $\gamma_0 \in C$ such that

$$f_{\gamma_0}(\omega) \leq f_\gamma(\omega) + \frac{\varepsilon}{3} P(A) \qquad \text{a.s. } P \tag{A.2.62}$$

$$f_{\gamma_0}(\omega) \leq f_{\gamma_n}(\omega) + \frac{\varepsilon}{3} P(A) \qquad \text{a.s. } P \tag{A.2.63}$$

$n = 1,2,\ldots,$. From (A.2.59) and (A.2.63)

$$f_{\gamma_0}(\omega) \leq g(\omega) + \frac{\varepsilon}{3} P(A) \qquad \text{a.s. } P . \tag{A.2.64}$$

Thus from (A.2.62) and (A.2.61)

$$\int_A f_{\gamma_0}(\omega)dP \leq \int_A f_\gamma(\omega)dP + \frac{\varepsilon}{3} P(A) \leq \int_A g(\omega)dP - \varepsilon P(A) + \frac{\varepsilon}{3} P(A) \quad . \tag{A.2.65}$$

From (A.2.64)

(A.2.66) $$\int_{A^c} f_{\gamma_0}(d)dP \leq \int_{A^c} g(\omega)dP + \frac{\epsilon}{3} P(A) \quad .$$

Adding (A.2.65) and (A.2.66)

$$\int_{\Omega} f_{\gamma_0}(\omega)dP \leq \int_{\Omega} g(\omega)dP - \frac{\epsilon}{3} P(A) \quad .$$

From (A.2.60)

$$\inf_{\gamma \in C} \int_{\Omega} f_{\gamma}(\omega)dP \leq \int_{\Omega} f_{\gamma_0}(\omega)dP \leq \int_{\Omega} g(\omega)dP - \frac{\epsilon}{3} P(A)$$

$$\leq \inf_{\gamma \in C} \int_{\Omega} f_{\gamma}(\omega)dP - \frac{\epsilon}{3} P(A) \quad .$$

Thus $P(A) = 0$ and it follows that

$$g(\omega) \leq f_{\gamma}(\omega) \qquad \text{a.s. } P \; .$$

This holds for all $\gamma \in C$, and from (A.2.59) $g(\omega)$ is measurable. Thus from iii) of Definition A.2.1

(A.2.67) $$g(\omega) \leq \operatorname*{P\text{-}ess\,inf}_{\gamma \in C} f_{\gamma}(\omega) \qquad \text{a.s. } P \; .$$

For $\epsilon > 0$, take $\gamma_0 \in C$ such that

$$f_{\gamma_0}(\omega) \leq f_{\gamma_n}(\omega) + \epsilon \qquad \text{a.s. } P$$

for $n = 1,2,\ldots$. Then from (A.2.59)

(A.2.68) $$f_{\gamma_0}(\omega) \leq g(\omega) + \epsilon \qquad \text{a.s. } P \; .$$

Finally, from ii) of Definition A.2.1 , (A.2.68), and (A.2.67)

(A.2.69) $$\operatorname*{P\text{-}ess\,inf}_{\gamma \in C} f_{\gamma}(\omega) \leq f_{\gamma_0}(\omega) \leq g(\omega) + \epsilon \leq \operatorname*{P\text{-}ess\,inf}_{\gamma \in C} f_{\gamma}(\omega) + \epsilon \qquad \text{a.s. } P \; .$$

It follows then from (A.2.69) and (A.2.59) that (A.2.57) holds for $\{f_{\gamma}\}$ integrable .

If the $\{f_{\gamma}\}$ are not integrable, for each integer N and f_{γ}^{N} defined by (A.2.20) take $\gamma_{N,n}$ such that

$$\inf_n f^N_{\gamma_{N,n}}(\omega) = \text{P-ess}\inf_{\gamma\in C} f^N_\gamma \qquad \text{a.s. } P .$$

Let

$$g(\omega) = \inf_{n,N} f_{\gamma_{N,n}}(\omega) ,$$

and g^N be defined by the truncation (A.2.20). Then

$$g^N(\omega) \leq \inf_n f^N_{\gamma_{N,n}}(\omega) = \text{P-ess}\inf_{\gamma\in C} f^N_\gamma \qquad \text{a.s. } P .$$

Letting $N \to \infty$, from Lemma A.2.5

$$g(\omega) \leq \text{P-ess}\inf_{\gamma\in C} f_\gamma \qquad \text{a.s. } P .$$

The argument is concluded as in the argument above following (A.2.67) thru (A.2.69).

<u>Lemma</u> A.2.16. Let $\{f_\gamma(\omega)\}$, $\gamma \in C$, be a family of measurable, a.s. non-negative (possibly $+\infty$) functions on $(\Omega,\mathfrak{A},P)$ with the countable ε-lattice property. Then for $\varepsilon > 0$, there exists $\gamma_0 \in C$ such that

$$\text{(A.2.70)} \qquad f_{\gamma_0}(\omega) - \varepsilon \leq \text{P-ess}\inf_{\gamma\in C} f_\gamma(\omega) \qquad \text{a.s. } P .$$

Proof: Let γ_n satisfy (A.2.57) of Lemma A.2.15. Then for $\varepsilon > 0$ from the countable ε-lattice property there exists $\gamma_0 \in C$ such that

$$f_{\gamma_0}(\omega) \leq f_{\gamma_n}(\omega) + \varepsilon \qquad \text{a.s. } P$$

for $n = 1,2,\ldots$. Thus

$$f_{\gamma_0}(\omega) - \varepsilon \leq \inf_n f_{\gamma_n}(\omega) \qquad \text{a.s. } P ,$$

and (A.2.70) follows from (A.2.57) .

<u>Theorem</u> A.2.3. Let $\{f_\gamma(\omega)\}$, $\gamma \in C$ be a family of measurable, a.s. non-negative (possibly $+\infty$) functions on $(\Omega,\mathfrak{A},P)$, let the family have the countable ε-lattice property, let $\mathfrak{B} \subset \mathfrak{A}$, and $P_{\mathfrak{B}}$ be the restriction of P to the σ-field $\mathfrak{B}$. Then (A.2.55) holds.

Proof: It can easily be shown that the family $\{E^{\mathcal{B}}[f_\gamma(\omega)]\}$, $\gamma \in C$ has the countable ε-lattice property. Thus from Lemma A.2.15 take $\gamma_n \in C$ and $\gamma'_n \in C$ such that

$$\inf_n f_{\gamma_n}(\omega) = \text{P-ess}\inf_{\gamma \in C} f_\gamma(\omega) \qquad \text{a.s. } P$$

$$\inf_n E^{\mathcal{B}} f_{\gamma'_n}(\omega) = P_{\mathcal{B}}\text{-ess}\inf_{\gamma \in C} E^{\mathcal{B}} f_\gamma(\omega) \qquad \text{a.s. } P_{\mathcal{B}} ,$$

and let

$$g(\omega) = \inf_n \min [f_{\gamma_n}(\omega) , f_{\gamma'_n}(\omega)] \quad .$$

Then

$$g(\omega) \leq f_{\gamma'_n}(\omega)$$

and

$$E^{\mathcal{B}}[g] \leq E^{\mathcal{B}}[f_{\gamma'_n}] \qquad \text{a.s. } P .$$

Thus

$$E^{\mathcal{B}}[g] \leq \inf_n E^{\mathcal{B}}[f_{\gamma'_n}] = P_{\mathcal{B}}\text{-ess}\inf_{\gamma \in C} E^{\mathcal{B}}[f_\gamma(\omega)] \qquad \text{a.s. } P , \tag{A.2.71}$$

and

$$g(\omega) \leq \inf_n f_{\gamma_n}(\omega) = \text{P-ess}\inf_{\gamma \in C} f_\gamma(\omega) \qquad \text{a.s. } P . \tag{A.2.72}$$

For $\varepsilon > 0$, from the countable ε-lattice property there exists γ_0 such that

$$f_{\gamma_0}(\omega) \leq f_{\gamma_n}(\omega) + \varepsilon \qquad \text{a.s. } P ,$$

$$f_{\gamma_0}(\omega) \leq f_{\gamma'_n}(\omega) + \varepsilon \qquad \text{a.s. } P .$$

Thus

$$f_{\gamma_0}(\omega) \leq g(\omega) + \varepsilon \qquad \text{a.s. } P ,$$

and hence

$$E^{\mathcal{B}}[f_{\gamma_0}(\omega)] \leq E^{\mathcal{B}}[g(\omega)] + \varepsilon \qquad \text{a.s. } P .$$

Thus from (A.2.71) and ii) of Definition A.2.1 for the family $\{E^{\mathcal{B}}[f_\gamma]\}$

$$E^{\mathcal{B}}[g] \leq P_{\mathcal{B}}\text{-ess}\inf_{\gamma \in C} E^{\mathcal{B}}[f_\gamma(\omega)] \leq E^{\mathcal{B}}[f_{\gamma_0}] \leq E^{\mathcal{B}}[g] + \varepsilon \qquad \text{a.s. } P .$$

Similarly from (A.2.72)

$$E^{\mathfrak{B}}[g] \leq E^{\mathfrak{B}}[\text{P-ess}\inf_{\gamma \in C} f_{\gamma}(\omega)] \leq E^{\mathfrak{B}}[f_{\gamma_0}] \leq E^{\mathfrak{B}}[g] + \epsilon \qquad \text{a.s. } P \ .$$

It follows then that

$$E^{\mathfrak{B}}[g] = P_{\mathfrak{B}}\text{-ess}\inf_{\gamma \in C} E^{\mathfrak{B}}[f_{\gamma}] = E^{\mathfrak{B}}[\text{P-ess}\inf_{\gamma \in C} f_{\gamma}] \qquad \text{a.s. } P \ .$$

<u>Lemma</u> A.2.17. Let $\{f_{\gamma}(\omega)\}$, $\gamma \in C$, be a family of measurable, a.s. non-negative (possibly $+\infty$) functions on $(\Omega, \mathfrak{A}, P)$, and let the family have the finite ϵ-lattice property and contain at least one integrable member. Then for $\gamma_0 \in C$

$$f_{\gamma_0}(\omega) = \text{P-ess}\inf_{\gamma \in C} f_{\gamma}(\omega) \qquad \text{a.s. } P \tag{A.2.73}$$

if and only if

$$\int f_{\gamma_0}(\omega)dP = \inf_{\gamma \in C} \int f_{\gamma}(\omega)dP \ . \tag{A.2.74}$$

Proof: If (A.2.73) holds, then

$$f_{\gamma_0}(\omega) \leq f_{\gamma}(\omega) \qquad \text{a.s. } P$$

for all $\gamma \in C$. Thus

$$\int f_{\gamma_0}(\omega)dP \leq \int f_{\gamma}(\omega)dP$$

for all $\gamma \in C$ and (A.2.74) follows since $\gamma_0 \in C$. Assume then that (A.2.74) holds. From Lemma A.2.14

$$\nu_{\gamma_0}(\Omega) = \nu(\Omega) < \infty \tag{A.2.75}$$

where ν and ν_{γ} are defined by (A.2.3) and (A.2.14). From (A.2.3)

$$\nu_{\gamma_0}(A) \leq \nu(A)$$

for all $A \in \mathfrak{A}$. Suppose there exists $A \in \mathfrak{A}$ such that

$$\nu_{\gamma_0}(A) < \nu(A) \ .$$

Then

$$\nu_{\gamma_0}(\Omega) = \nu_{\gamma_0}(A) + \nu_{\gamma_0}(A^c) < \nu(A) + \nu(A^c) = \nu(\Omega)$$

contradicting (A.2.75). Thus

(A.2.76) $$\nu_{\gamma_0}(A) = \nu(A)$$

for all $A \in \mathfrak{A}$. From (A.2.14), (A.2.76), and (A.2.15) of Lemma A.2.4

$$f_{\gamma_0}(\omega) = \frac{d\nu_{\gamma_0}}{dP} = \frac{d\nu}{dP} = \text{P-ess}\inf_{\gamma \in C} f_\gamma(\omega) \qquad \text{a.s. } P .$$

Lemma A.2.18. Let $\{f_\gamma\}$, $\gamma \in C$, be a family of measurable, a.s. non-negative (possibly $+\infty$) functions on $(\Omega,\mathfrak{A},P)$, and let $f(\omega)$ be a measurable function that satisfies

(A.2.77) $$f(\omega) = \inf_\gamma f_\gamma(\omega) \qquad \text{a.s. } P .$$

Then

(A.2.78) $$f(\omega) \leq \text{P-ess}\inf_\gamma f_\gamma(\omega) \qquad \text{a.s. } P .$$

Proof: From (A.2.77)

$$f(\omega) \leq f_\gamma(\omega) \qquad \text{a.s. } P$$

for all $\gamma \in C$. Thus (A.2.78) follows from iii) of Definition A.2.1.

The following example shows that (A.2.77) strengthened by the countable ϵ-lattice property still does not imply that (A.2.78) holds with a.s. equality.

Let $\{N_\gamma\}$, $\gamma \in C$ be the class of all countable subsets of the unit interval $[0,1]$ and define $f_\gamma(\omega)$ by

$$f_\gamma(\omega) = \begin{cases} 1 & \omega \notin N_\gamma \\ 0 & \omega \in N_\gamma \end{cases} .$$

Then for P Lebesgue measure on $[0,1]$ clearly

$$\text{P-ess}\inf_{\gamma \in C} f_\gamma(\omega) = 1 \qquad \text{a.s. } P ,$$

and

$$\inf_{\gamma} f_\gamma(\omega) = 0 \qquad \text{for all } \omega \in [0,1] .$$

The family $\{f_\gamma(\omega)\}$ has the countable ϵ-lattice property.

<u>Lemma</u> A.2.19. Let $\{f_\gamma(\omega)\}$, $\gamma \in C$, be a family of measurable a.s. non-negative (possibly $+\infty$) functions on $(\Omega,\mathfrak{A},P)$, and let the family have the property that for each $\epsilon > 0$, there exists $\gamma_\epsilon \in C$ such that

(A.2.79) $$f_{\gamma_\epsilon}(\omega) \leq \inf_{\gamma \in C} f_\gamma(\omega) + \epsilon \qquad \text{a.s. } P .$$

Then

(A.2.80) $$\inf_{\gamma \in C} f_\gamma(\omega) = P\text{-ess}\inf_{\gamma \in C} f_\gamma(\omega) \qquad \text{a.s. } P .$$

Proof: Let $\gamma_n \in C$ satisfy

(A.2.81) $$f_{\gamma_n}(\omega) \leq \inf_{\gamma \in C} f_\gamma(\omega) + \frac{1}{n} \qquad \text{a.s. } P$$

for $n = 1,2,\ldots,$, and let

(A.2.82) $$f(\omega) = \inf_n f_{\gamma_n}(\omega) .$$

Since the $f_{\gamma_n}(\omega)$ are measurable, $f(\omega)$ is measurable. From (A.2.81) and (A.2.82)

$$f(\omega) \leq f_{\gamma_n}(\omega) \leq \inf_{\gamma \in C} f_\gamma(\omega) + \frac{1}{n} \qquad \text{a.s. } P$$

for all n . Thus

$$f(\omega) \leq \inf_{\gamma \in C} f_\gamma(\omega) \qquad \text{a.s. } P .$$

Since $\gamma_n \in C$, for all ω

$$\inf_{\gamma \in C} f_\gamma(\omega) \leq \inf_n f_{\gamma_n}(\omega) = f(\omega) .$$

Thus

$$f(\omega) = \inf_{\gamma \in C} f_\gamma(\omega) \qquad \text{a.s. } P .$$

From ii) of Definition A.2.1

$$P\text{-ess}\inf_{\gamma \in C} f_\gamma(\omega) \leq f_{\gamma_n}(\omega) \qquad \text{a.s. } P$$

for all n . Thus

$$P\text{-ess}\inf_{\gamma\in C} f_\gamma(\omega) \le \inf_n f_{\gamma_n}(\omega) = f(\omega) = \inf_{\gamma\in C} f_\gamma(\omega) \quad \text{a.s. } P\ ,$$

and (A.2.80) follows from Lemma A.2.18 .

References

[1] Aoki, Masanao, Opimization of Stochastic Systems, Academic Press, New York, 1967.

[2] Blackwell, David, "Discounted dynamic programming," Annals of Math. Stat., Vol 36, pp. 226-235 (1965).

[3] Brown, L.D. and Purves, R. "Measurable selections of extrema" Annals of Stat., Vol. 1, pp. 902-912 (1973) .

[4] Doob, J.L., Stochastic Processes, John Wiley and Sons, Inc., New York, 1953.

[5] Dubins, Lester E. and Savage, Leonard J., How To Gamble If You Must, McGraw-Hill, New York (1965).

[6] Dunford, Nelson and Schwartz, Jacob T., Linear Operators, Part I: General Theory, Interscience Publishers. Inc., New York (1957).

[7] Dynkin, E.B., "Controlled random processes," Theory of Probability and its Applications (translation), Vol. 10, pp. 1-14 (1965).

[8] Hinderer, K., Foundations of Non-stationary Dynamic Programming with Discrete Time Parameter, Lecture Notes in Operations Research and Mathematical Systems, No. 33, Springer-Verlag, Berlin, 1970.

[9] Kalman, R.E., "New methods and results in linear prediction and filtering theory,"Research Institute for Advanced Studies, Rept.61-1, Martin Co., Baltimore, Md. (1960).

[10] Karlin, Samuel, Total Positivity, Vol. 1, Stanford University Press, Stanford, Calif., 1968.

[11] Lehmann, E.L., Testing Statistical Hypotheses, John Wiley and Sons, Inc., New York, 1959.

[12] Loève, Michel, Probability Theory, 3rd ed., D. van Nostrand Co., Princeton, New Jersey, 1963.

[13] Maitra, A. "Discounted dynamic programming on compact metric spaces," Sankhya, Vol. 30A, p. 211-219 (1968).

[14] Rauch, H.E., Tung, F., and Striebel, C.T., "Maximum likelihood estimates of linear dynamic systems," AIAA Journal, Vol.3, pp. 1445-1450 (1965).

[15] Sirjaev, A.N. "Some new results in the theory of controlled random processes," Selected Translations in Mathematical Statistics and Probability, Vol. 8 pp. 49-130 (1970).

[16] Strauch, R.E., "Negative dynamic programming," Annals of Math. Stat. Vol. 37, pp. 871-890 (1966).

[17] Striebel, Charlotte, "Sufficient statistics in the optimal control of stochastic systems," Journal of Math. Anal. and Appl., Vol. 12, pp. 576-592, (1965).

[18] Striebel, Charlotte, "Martingale conditions for the optimal control of continuous time stochastic systems," presented at the International Workshop on Stochastic Filtering and Control, Los Angeles, Calif., May 13-17, 1974.

[19] Tung, F. and Striebel, C.T., "A stochastic optimal control problem and its applications," _Journal of Math. Anal. and Appl._, Vol. 12, pp. 350-360. (1965).

[20] Witsenhausen, Hans S., "Separation of estimation and control for discrete time systems," _Proceedings of the IEEE_, Vol. 59, pp. 1557-1566 (1971).

Vol. 59: J. A. Hanson, Growth in Open Economies. V, 128 pages. 1971. DM 18,–

Vol. 60: H. Hauptmann, Schätz- und Kontrolltheorie in stetigen dynamischen Wirtschaftsmodellen. V, 104 Seiten. 1971. DM 18,–

Vol. 61: K. H. F. Meyer, Wartesysteme mit variabler Bearbeitungsrate. VII, 314 Seiten. 1971. DM 27,–

Vol. 62: W. Krelle u. G. Gabisch unter Mitarbeit von J. Burgermeister, Wachstumstheorie. VII, 223 Seiten. 1972. DM 22,–

Vol. 63: J. Kohlas, Monte Carlo Simulation im Operations Research. VI, 162 Seiten. 1972. DM 18,–

Vol. 64: P. Gessner u. K. Spremann, Optimierung in Funktionenräumen. IV, 120 Seiten. 1972. DM 18,–

Vol. 65: W. Everling, Exercises in Computer Systems Analysis. VIII, 184 pages. 1972. DM 20,–

Vol. 66: F. Bauer, P. Garabedian and D. Korn, Supercritical Wing Sections. V, 211 pages. 1972. DM 22,–

Vol. 67: I. V. Girsanov, Lectures on Mathematical Theory of Extremum Problems. V, 136 pages. 1972. DM 18,–

Vol. 68: J. Loeckx, Computability and Decidability. An Introduction for Students of Computer Science. VI, 76 pages. 1972. DM 18,–

Vol. 69: S. Ashour, Sequencing Theory. V, 133 pages. 1972. DM 18,–

Vol. 70: J. P. Brown, The Economic Effects of Floods. Investigations of a Stochastic Model of Rational Investment. Behavior in the Face of Floods. V, 87 pages. 1972. DM 18,–

Vol. 71: R. Henn und O. Opitz, Konsum- und Produktionstheorie II. V, 134 Seiten. 1972. DM 18,–

Vol. 72: T. P. Bagchi and J. G. C. Templeton, Numerical Methods in Markov Chains and Bulk Queues. XI, 89 pages. 1972. DM 18,–

Vol. 73: H. Kiendl, Suboptimale Regler mit abschnittweise linearer Struktur. VI, 146 Seiten. 1972. DM 18,–

Vol. 74: F. Pokropp, Aggregation von Produktionsfunktionen. VI, 107 Seiten. 1972. DM 18,–

Vol. 75: GI-Gesellschaft für Informatik e.V. Bericht Nr. 3. 1. Fachtagung über Programmiersprachen · München, 9.–11. März 1971. Herausgegeben im Auftrag der Gesellschaft für Informatik von H. Langmaack und M. Paul. VII, 280 Seiten. 1972. DM 27,–

Vol. 76: G. Fandel, Optimale Entscheidung bei mehrfacher Zielsetzung. 121 Seiten. 1972. DM 18,–

Vol. 77: A. Auslender, Problèmes de Minimax via l'Analyse Convexe et les Inégalités Variationelles: Théorie et Algorithmes. VII, 132 pages. 1972. DM 18,–

Vol. 78: GI-Gesellschaft für Informatik e.V. 2. Jahrestagung, Karlsruhe, 2.–4. Oktober 1972. Herausgegeben im Auftrag der Gesellschaft für Informatik von P. Deussen. XI, 576 Seiten. 1973. DM 40,–

Vol. 79: A. Berman, Cones, Matrices and Mathematical Programming. V, 96 pages. 1973. DM 18,–

Vol. 80: International Seminar on Trends in Mathematical Modelling, Venice, 13–18 December 1971. Edited by N. Hawkes. VI, 288 pages. 1973. DM 27,–

Vol. 81: Advanced Course on Software Engineering. Edited by F. L. Bauer. XII, 545 pages. 1973. DM 35,–

Vol. 82: R. Saeks, Resolution Space, Operators and Systems. X, 267 pages. 1973. DM 24,–

Vol. 83: NTG/GI-Gesellschaft für Informatik, Nachrichtentechnische Gesellschaft. Fachtagung „Cognitive Verfahren und Systeme", Hamburg, 11.–13. April 1973. Herausgegeben im Auftrag der NTG/GI von Th. Einsele, W. Giloi und H.-H. Nagel. VIII, 373 Seiten. 1973. DM 32,–

Vol. 84: A. V. Balakrishnan, Stochastic Differential Systems I. Filtering and Control. A Function Space Approach. V, 252 pages. 1973. DM 24,–

Vol. 85: T. Page, Economics of Involuntary Transfers: A Unified Approach to Pollution and Congestion Externalities. XI, 159 pages. 1973. DM 20,–

Vol. 86: Symposium on the Theory of Scheduling and Its Applications. Edited by S. E. Elmaghraby. VIII, 437 pages. 1973. DM 35,–

Vol. 87: G. F. Newell, Approximate Stochastic Behavior of n-Server Service Systems with Large n. VIII, 118 pages. 1973. DM 18,–

Vol. 88: H. Steckhan, Güterströme in Netzen. VII, 134 Seiten. 1973. DM 18,–

Vol. 89: J. P. Wallace and A. Sherret, Estimation of Product. Attributes and Their Importances. V, 94 pages. 1973. DM 18,–

Vol. 90: J.-F. Richard, Posterior and Predictive Densities for Simultaneous Equation Models. VI, 226 pages. 1973. DM 22,–

Vol. 91: Th. Marschak and R. Selten, General Equilibrium with Price-Making Firms. XI, 246 pages. 1974. DM 24,–

Vol. 92: E. Dierker, Topological Methods in Walrasian Economics. IV, 130 pages. 1974. DM 18,–

Vol. 93: 4th IFAC/IFIP International Conference on Digital Computer Applications to Process Control, Zürich/Switzerland, March 19–22, 1974. Edited by M. Mansour and W. Schaufelberger. XVIII, 544 pages. 1974. DM 40,–

Vol. 94: 4th IFAC/IFIP International Conference on Digital Computer Applications to Process Control, Zürich/Switzerland, March 19–22, 1974. Edited by M. Mansour and W. Schaufelberger. XVIII, 546 pages. 1974. DM 40,–

Vol. 95: M. Zeleny, Linear Multiobjective Programming. XII, 220 pages. 1974. DM 22,–

Vol. 96: O. Moeschlin, Zur Theorie von Neumannscher Wachstumsmodelle. XI, 115 Seiten. 1974. DM 16,–

Vol. 97: G. Schmidt, Über die Stabilität des einfachen Bedienungskanals. VII, 147 Seiten. 1974. DM 16,–

Vol. 98: Mathematical Methods in Queueing Theory. Proceedings of a Conference at Western Michigan University, May 10–12, 1973. Edited by A. B. Clarke. VII, 374 pages. 1974. DM 28,–

Vol. 99: Production Theory. Edited by W. Eichhorn, R. Henn, O. Opitz, and R. W. Shephard. VIII, 386 pages. 1974. DM 32,–

Vol. 100: B. S. Duran and P. L. Odell, Cluster Analysis. A survey. VI, 137 pages. 1974. DM 18,–

Vol. 101: W. M. Wonham, Linear Multivariable Control. A Geometric Approach. X, 344 pages. 1974. DM 30,–

Vol. 102: Analyse Convexe et Ses Applications. Comptes Rendus, Janvier 1974. Edited by J.-P. Aubin. IV, 244 pages. 1974. DM 25,–

Vol. 103: D. E. Boyce, A. Farhi, R. Weischedel, Optimal Subset Selection. Multiple Regression, Interdependence and Optimal Network Algorithms. XIII, 187 pages. 1974. DM 20,–

Vol. 104: S. Fujino, A Neo-Keynesian Theory of Inflation and Economic Growth. V, 96 pages. 1974. DM 18,–

Vol. 105: Optimal Control Theory and its Applications. Part I. Proceedings of the Fourteenth Biennual Seminar of the Canadian Mathematical Congress. University of Western Ontario, August 12–25, 1973. Edited by B. J. Kirby. VI, 425 pages. 1974. DM 35,–

Vol. 106: Optimal Control Theory and its Applications. Part II. Proceedings of the Fourteenth Biennial Seminar of the Canadian Mathematical Congress. University of Western Ontario, August 12-25, 1973. Edited by B. J. Kirby. VI, 403 pages. 1974. DM 35,–

Vol. 107: Control Theory, Numerical Methods and Computer Systems Modelling. International Symposium, Rocquencourt, June 17–21, 1974. Edited by A. Bensoussan and J. L. Lions. VIII, 757 pages. 1975. DM 53,–

Vol. 108: F. Bauer et al., Supercritical Wing Sections II. A Handbook. V, 296 pages. 1975. DM 28,–

Vol. 109: R. von Randow, Introduction to the Theory of Matroids. IX, 102 pages. 1975. DM 18,–

Vol. 110: C. Striebel, Optimal Control of Discrete Time Stochastic Systems. III. 208 pages. 1975. DM 23,–